CHINA'S FUTURE:

Foreign Policy and Economic Development in the Post-Mao Era

Studies by Allen S. Whiting and by Robert F. Dernberger

ALTERNATIVES TO MONETARY DISORDER

Studies by Fred Hirsch and Michael W. Doyle and by Edward L. Morse

NUCLEAR PROLIFERATION:

Motivations, Capabilities, and Strategies for Control

Studies by Ted Greenwood and by Harold A. Feiveson and Theodore B. Taylor

INTERNATIONAL DISASTER RELIEF:

Toward a Responsive System

Stephen Green

STUDIES FORTHCOMING

The 1980s Project will comprise about 30 volumes. Most will contain independent but related studies concerning issues of potentially great importance in the next decade and beyond, such as resource management, human rights, population studies, and relations between the developing and developed societies, among many others. Additionally, a number of volumes will be devoted to particular regions of the world, concentrating especially on political and economic development trends outside the industrialized West.

Oil Politics in the 1980s

PATTERNS OF INTERNATIONAL COOPERATION

ØYSTEIN NORENG

1980s Project/Council on Foreign Relations

McGRAW-HILL BOOK COMPANY

New York   St. Louis   San Francisco
Auckland   Bogotá   Düsseldorf   Johannesburg   London   Madrid
Mexico   Montreal   New Delhi   Panama   Paris   São Paulo
Singapore   Sydney   Tokyo   Toronto

The Council on Foreign Relations, Inc. is a nonprofit and nonpartisan organization devoted to promoting improved understanding of international affairs through the free exchange of ideas. Its membership of about 1,700 persons throughout the United States is made up of individuals with special interest and experience in international affairs. The Council has no affiliation with and receives no funding from the United States government.

The Council publishes the quarterly journal *Foreign Affairs* and, from time to time, books and monographs which in the judgment of the Council's Committee on Studies are responsible treatments of significant international topics worthy of presentation to the public. The 1980s Project is a research effort of the Council; as such, 1980s Project Studies have been similarly reviewed through procedures of the Committee on Studies. As in the case of all Council publications, statements of fact and expressions of opinion contained in 1980s Project Studies are the sole responsibility of their authors.

The editor of this book was Thomas Wallin for the Council on
Foreign Relations. Thomas Quinn and Michael Hennelly were the editors
for McGraw-Hill Book Company. Christopher Simon was the designer and
Teresa Leaden supervised the production. This book was set in
Times Roman by Creative Book Services, Inc.

Printed and bound by R. R. Donnelley & Sons.

Library of Congress Cataloging in Publication Data

Noreng, Øystein
Oil politics in the 1980s.

(1980s project/Council on Foreign Relations)
Bibliography: p.
Includes index.
1. Petroleum industry and trade. 2. Petroleum
industry and trade—United States. 3. Organization of
Petroleum Exporting Countries. 4. International
economic relations. I. Title. II. Series: Council
on Foreign Relations. 1980s project/Council on
Foreign Relations.
HD9560.5.N63   382'.42'282   78-9134
ISBN 0-07-047185-1
ISBN 0-07-047186-X pbk.

1 2 3 4 5 6 7 8 9 R R D R R D 7 8 3 2 1 0 9 8

Contents

Foreword: The 1980s Project

The study in this volume, proposing the framework of a negotiated agreement between the major exporters and importers of petroleum, is part of a stream of studies commissioned by the 1980s Project of the Council on Foreign Relations. Like other 1980s Project studies, it analyzes an issue that is likely to be of major international concern during the coming decade or two.

The ambitious purpose of the 1980s Project is to examine important political and economic problems not only individually but in relationship to one another. Some studies or books produced by the Project will primarily emphasize the interrelationship of issues. In the case of other, more specifically focused studies, a considerable effort has been made to write, review, and criticize them in the context of more general Project work. Each Project study is thus capable of standing on its own; at the same time it has been shaped by a broader perspective.

This volume on the politics of petroleum, like most other 1980s Project studies, attempts to transcend artificial disciplinary boundaries between politics and economics. As well, like other volumes in this series, the author's analysis is essentially normative and prescriptive. The thrust of Øystein Noreng's argument is that in the regulation of the international oil markets a commonality of interests exists between major importing and exporting countries, concerning both the relative price of oil and the need to regulate oil supply on a guaranteed basis. He puts forward proposals that would enable both exporting and importing countries to make more certain that their commonality of interests will override their inevitable conflicts of interests. And in putting forward his recommendations, he analyzes concensus and dissension among both OPEC members and countries in the industrial West.

The author brings to his analysis and to his policy recommendations a perspective derived from his own personal and national experiences. A trained political scientist who received his education in France and who has carried on research in the United States, and a former official of the Ministry of Finance of the Norwegian government and of the Norwegian state oil company, Statoil, Øystein Noreng is acutely sensitive to the perspective of officials in both oil-importing and -exporting countries. As a European, he brings to bear a special sensitivity to the ways in which American domestic energy policies and the size of America's petroleum imports affect other industrialized countries. As a citizen of a small exporting country within the industrial world, he is alert to possibilities for compromise between OPEC countries and OECD importing countries. And as a student of politics, he acknowledges the very special roles being played in the world oil market by the United States—the world's largest petroleum importer and most industrialized country—and Saudi Arabia—the world's largest residual producer of petroleum.

The recommendations put forward in this volume for an accommodation between the world's major oil-exporting and -importing countries is similar in some respects to proposals tabled at the Conference on International Economic Cooperation—the "North/South Conference"—held in Paris from 1975 to 1977. Many persons argued then and continue to do so now that international institutional arrangements are not desirable or feasible in this area. The Noreng study is premised on the notion that conditions for an international agreement have changed substantially since then and that an increased understanding of some major issues related to such an agreement will make it more feasible in coming years. For example, knowledge of world oil reserves, of the elasticity of demand for oil in the United States, and of the potential of oil-exporting countries for investing in industrial countries has expanded dramatically since the mid-1970s. Moreover, one of the noteworthy aspects of the proposals put forward by Øystein Noreng is that they flow logically from a

European perspective on world petroleum politics, and it was some European governments together with OPEC that stood in the way of similar recommendations a few years ago.

Although this study on the world politics of petroleum is relatively self-contained, it raises a wide range of issues that bear upon other central features of future world politics. For example, efforts rationally to plan and manage world oil supplies and prices and to diffuse the sometimes bitter politics associated with them are suggestive of ways that other commodities ought—or ought not—to be handled. As well, given the central role played by OPEC nations as upper-income developing countries in the North-South dialogue, whether a global oil agreement can be reached—and the manner by which it might be reached—will affect the level of confrontation or accommodation as well as bargaining relationships over such issues as equity in the sharing of world income and resources and trade relationships between rich and poor countries. Similarly, the balance-of-payments surpluses and deficits associated with the oil-price increases of 1973–1974 have placed heavy burdens on the international financial system and have altered the general pattern of global growth and investment. Finally, the issue of petroleum itself and the way problems associated with it are resolved will directly influence the development of new sources of petroleum and other sources of energy.

The 1980s Project had its origins in the widely held recognition that many of the assumptions, policies, and institutions that have characterized international relations during the past 30 years are inadequate to the demands of today and the foreseeable demands of the period between now and 1990 or so. Over the course of the next decade, substantial adaptation of institutions and behavior will be needed to respond to the changed circumstances of the 1980s and beyond. The Project seeks to identify those future conditions and the kinds of adaptation they might require. It is not the Project's purpose to arrive at a single or exclusive set of goals. Nor does it focus upon the foreign policy or national interests of the United States alone. Instead, it seeks to identify goals that are compatible with the perceived interests of most states, despite differences in ideology and in level of economic development.

The published products of the Project are aimed at a broad

readership, including policy makers and potential policy makers and those who would influence the policy-making process, but are confined to no single nation or region. The authors of Project studies were therefore asked to remain mindful of interests broader than those of any one society and to take fully into account the likely realities of domestic politics in the principal societies involved. All those who have worked on the Project, however, have tried not to be captives of the status quo; they have sought to question the inevitability of existing patterns of thought and behavior that restrain desirable change and to look for ways in which those patterns might in time be altered or their consequences mitigated.

The 1980s Project is at once a series of separate attacks upon a number of urgent and potentially urgent international problems and also a collective effort, involving a substantial number of persons in the United States and abroad, to bring those separate approaches to bear upon one another and to suggest the kinds of choices that might be made among them. The Project involves more than 300 participants. A small central staff and a steering Coordinating Group have worked to define the questions and to assess the compatibility of policy prescriptions. Nearly 100 authors, from more than a dozen countries, have been at work on separate studies. Ten working groups of specialists and generalists have been convened to subject the Project's studies to critical scrutiny and help in the process of identifying interrelationships among them.

The 1980s Project is the largest single research and studies effort the Council on Foreign Relations has undertaken in its 55-year history, comparable in conception only to a major study of the postwar world, the War and Peace Studies, undertaken by the Council during the Second World War. At that time, the impetus to the effort was the discontinuity caused by worldwide conflict and the visible and inescapable need to rethink, replace, and supplement many of the features of the international system that had prevailed before the war. The discontinuities in today's world are less obvious and, even when occasionally quite visible—as in the abandonment of gold convertibility and fixed monetary parities—only briefly command the spotlight of public attention. That new institutions and patterns of behavior are needed in many areas is widely acknowledged, but the sense of need is less urgent—existing institutions have not for the most part dramatically failed and collapsed. The tendency, therefore, is to make do with outmoded arrangements and to improvise rather than to undertake a basic analysis of the problems that lie before us and of the demands that those problems will place upon all nations.

The 1980s Project is based upon the belief that serious effort and integrated forethought can contribute—indeed, are indispensable—to progress in the next decade toward a more humane, peaceful, productive, and just world. And it rests upon the hope that participants in its deliberations and readers of Project publications—whether or not they agree with an author's point of view—may be helped to think more informedly about the opportunities and the dangers that lie ahead and the consequences of various possible courses of future action.

The 1980s Project has been made possible by generous grants from the Ford Foundation, the Lilly Endowment, the Andrew W. Mellon Foundation, the Rockefeller Foundation, and the German Marshall Fund of the United States. Neither the Council on Foreign Relations nor any of those foundations is responsible for statements of fact and expressions of opinion contained in publications of the 1980s Project; they are the sole responsibility of the individual authors under whose names they appear. But the Council on Foreign Relations and the staff of the 1980s Project take great pleasure in placing those publications before a wide readership both in the United States and abroad.

Catherine Gwin
Edward L. Morse
Richard H. Ullman

1980s PROJECT WORKING GROUPS

During 1975 and 1976, ten Working Groups met to explore major international issues and to subject initial drafts of 1980s Project studies to critical review. Those who chaired Project Working Groups were:

Cyrus R. Vance, Working Group on Nuclear Weapons and Other Weapons of Mass Destruction.

Leslie H. Gelb, Working Group on Armed Conflict

Roger Fisher, Working Group on Transnational Violence and Subversion

Rev. Theodore M. Hesburgh, Working Group on Human Rights

Joseph S. Nye, Jr., Working Group on the Political Economy of North-South Relations

Harold Van B. Cleveland, Working Group on Macroeconomic Policies and International Monetary Relations

Lawrence C. McQuade, Working Group on Principles of International Trade

William Diebold, Jr., Working Group on Multinational Enterprises

Eugene B. Skolnikoff, Working Groups on the Environment, the Global Commons, and Economic Growth

Miriam Camps, Working Group on Industrial Policy

1980s PROJECT STAFF

Persons who have held senior professional positions on the staff of the 1980s Project for all or part of its duration are:

Miriam Camps	*Catherine Gwin*
William Diebold, Jr.	*Roger D. Hansen*
Tom J. Farer	*Edward L. Morse*
David C. Gompert	*Richard H. Ullman*

Richard H. Ullman was Director of the 1980s Project from its inception in 1974 until July 1977, when he became Chairman of the Project Coordinating Group. Edward L. Morse was Executive Director from July 1977 until June 1978. At that time, Catherine Gwin, 1980s Project Fellow since 1976, took over as Executive Director.

xiii

PROJECT COORDINATING GROUP

The Coordinating Group of the 1980s Project had a central advisory role in the work of the Project. Its members as of June 30, 1978, were:

Carlos F. Díaz-Alejandro   Bayless Manning
Richard A. Falk   Theodore R. Marmor
Tom J. Farer   Ali Mazrui
Edward K. Hamilton   Michael O'Neill
Stanley Hoffmann   Stephen Stamas
Gordon J. MacDonald   Fritz Stern
Bruce K. MacLaury   Allen S. Whiting

Until they entered government service, other members included:

W. Michael Blumenthal   Joseph S. Nye, Jr.
Richard N. Cooper   Marshall D. Schulman
Samuel P. Huntington

COMMITTEE ON STUDIES

The Committee on Studies of the Board of Directors of the Council on Foreign Relations is the governing body of the 1980s Project. The Committee's members as of June 30, 1978, were:

Barry E. Carter   Robert E. Osgood
Robert A. Charpie   Stephen Stamas
Stanley Hoffmann   Paul A. Volcker
Henry A. Kissinger   Marina v.N. Whitman
Walter J. Levy

James A. Perkins (Chairman)

Acknowledgments

This essay was written under quite changing circumstances, reflecting my own professional itinerary over the past years. Consequently I am obliged to a number of institutions and individuals. First, my gratitude goes to the 1980s Project Staff of the Council on Foreign Relations. When the Council commissioned me to write a paper on oil, I was working as a counselor in the planning department of the Norwegian Finance Ministry, where I had the pleasure of being aided and inspired by my colleagues, in particular Per Schreiner, Amund Utne, and Trond Reinertsen. From 1976 to 1977 I was research and planning manager in the marketing department of Statoil, Norway's state oil company. Several of my colleagues were of considerable assistance, in particular Jarle-Erik Sandvik, Bengt Ramberg, Nils Sletten, Per-Otto Skaug, and Erik Schanche. In 1977 I was appointed studies director (professor) at the Oslo Institute of Business Administration. I am particularly indebted to the Reter Garson Komissar and to the research committee for their generous support, as well as to my colleagues Pal Korsvold and Fred Wenstop for advice. Finally, the work was completed at Stanford University, where, on a Fulbright fellowship, I was a guest of the Institute for Energy Studies. I am particularly grateful to professors Holt Ashley, Tom Conolly, and William Reynolds as well as to Mrs. Marian Rees and the staff of the department of aeronautics and astronautics. My research assistant Warren Cordell was most helpful and efficient in preparing the scenarios.

Outside this circuit, a few other institutions and persons should be mentioned. The United States Educational Foundation in Norway, the Norwegian office of the First National City Bank, the Norway-American Association, and Mobil Exploration Norway provided travel grants that permitted me to make several visits to the United States. Helpful advice was provided by Joel Darmstadter of Resources for the Future, Professors Irma Adelman of the University of Maryland, Harvey Brooks and Eugene B. Skolnikoff of Harvard University, Dankwart Rustow at the City University of New York, as well as Richard Rudman of the Electric Power-Research Institute (EPRI), and Keith Pavitt of the University of Sussex. Mr. J. Alexander Caldwell, vice president of Crocker National Bank, was most helpful in providing data on OPEC. Also, Johan Nic. Vold of the Norwegian Ministry of Industry, Francisco Parra of Petroven, London, and Jean Carrié of the Compagnie Française de Pétrole, as well as Professor Jean-Marie Chevalier of the University of Grenoble, were, over a long period of time, most helpful in explaining to me the intricacies of the world oil market. Over many years, I have enjoyed the cooperation of the Norwegian Institute of International Affairs. I have had the pleasure of discussing oil and energy matters with the director, Professor John Sanness, and research directors, Daniel Heradstreit and Martin Saeter. During the last phase of the work, I had the pleasure of discussing the outlook for the world oil market with James Hambling and John Mitchell of British Petroleum. Dr. Mason Willrich, Director of International Relations at the Rockefeller Foundation, has given some useful comments in the final phase of the work. Finally, I have on several occasions been the guest of the International Research Center for Energy and Economic Development (ICEED) of the University of Colorado, enjoying the insight and inspiration of Professor Ragaei El Mallakh.

Given my background, readers might ask to what extent this essay represents a Norwegian point of view on the world's energy problems. The aim of the agreement proposed here is to reconcile the OECD and OPEC countries for the benefit of common economic growth. This proposal reflects Norway's position, at present the only net exporter of oil in the Western world, which gives it both OECD and OPEC interests. The Norwegian government has made clear, in important policy documents, its preference for international energy cooperation. However, the points of view and the ideas put forth in this essay are strictly my own and should not be attributed to the Norwegian government or to Statoil. Many of the points of view presented here might seem alien, if not provocative, to the American public. To a certain extent they reflect a Western European understanding of oil and energy problems and tend to underline the responsibility of the United States. My aim is to stimulate thought and discussion by outlining possible patterns of cooperation on problems that concern us all.

Øystein Noreng

ONE

An International Oil Agreement: Introduction

The energy crisis of the 1970s took most of the world by surprise. Studies of the future made in the 1960s hardly mention energy as a potential problem.[1] In the early 1970s, energy rather suddenly became a major concern for the world economy and a major issue in international politics. The prominence of energy is likely to continue, given the rather unexpected behavior of the international energy market in recent years. The elements of this surprising behavior include large price increases followed by reduced supply prospects and little significant change in the patterns of demand. This demonstrates the limited resource base for oil,[2] and also shows that the fairly close historical relationship between the level of energy consumption and the level of economic activity still is valid, even at much higher prices.[3] Actually, the reduction of energy consumption in most Western industrial countries in 1974 and 1975 was not due to the price increase as much as it was the result of the economic recession and, to a lesser extent, the mild winters. Although the energy sector accounts for only a small percentage of the gross national product of the industrialized consuming countries, energy is a

[1]See Herman Kahn and Anthony J. Wiener, *The Year 2000*, Macmillan, New York, 1967, p. 75ff. They hardly mention oil or energy as a potential problem, but assume an abundance in the near future.

[2]Nicholas Georgescu-Roegen, "Energy and Economic Myths," *Southern Economic Journal*, vol. 41, no. 3, January 1975, pp. 347–381.

[3]Neil H. Jacoby, *Multinational Oil*, Macmillan, New York, 1974, p. 15ff.

basic necessity for the functioning of any industrial economy.[4] Thus energy supply problems caused by physical shortages or high prices are likely to affect directly the economic growth of the industrialized consuming countries.

On the basis of this recent experience, it can be argued that market forces, particularly through the price mechanism, are only going to be effective in establishing an equilibrium in the energy market over a long period of time, perhaps 25 to 30 years. In the meantime, political factors such as the policies and preferences of a few key countries could have a profound influence upon the long-term "solutions" worked out by market forces. The response of energy demand to price changes seems to be of less importance and slower than previously anticipated,[5] and it also seems to be fairly limited, particularly outside North America. This is partly because the discovery rate for oil has fallen in recent years and partly because substitutes seem to be increasingly capital-intensive.

This limited ability of the price mechanism to regulate the demand and supply for energy creates the potential for abrupt changes, primarily price discontinuities. It is the impact of organizational and political factors that is critical in the world energy market. In this century, the price of oil has only to a limited extent been determined by open transactions in a free market.[6] In fact, there is historically no systematic relationship between the price of oil and the costs of production.[7] In the past, the key organizational and political factors worked to the advantage of the industrialized consumers. As control of the oil market has moved

[4]Philip Connelly and Robert Perlman, *The Politics of Scarcity: Resource Conflicts in International Relations*, Oxford University Press, London, 1975, p. 26ff.

[5]Joel Darmstadter, Joy Dunkerley, and Jack Alternan, *How Industrial Societies Use Energy: A Comparative Analysis*, Resources for the Future, Washington, D.C., 1977, p. 183ff.

[6]Robert Engler, *The Brotherhood of Oil*, University of Chicago Press, Chicago, 1977, p. 16ff.

[7]Douglas R. Bohl and Milton Russell, "Some Economic Effects of the United States Oil Import Quota," in Ragaei El Mallakh and Carl McGuire (eds.), *U.S. and World Energy Resources: Prospects and Priorities*, ICEED, Boulder, Colo., 1977, pp. 1–19.

2

to the producers, the industrialized consumers increasingly run the risk of new price rises and even supply crises. To avoid economic and political disturbances in the short run, and to facilitate the achievement of an equilibrium in the long run, it is in the interest of the industrialized consumers to find a political solution to the world's energy problem, in the form of a negotiated international agreement. In order to be viable, such an agreement must also secure vital interests for the producers.

The initial reaction to the fourfold increase in the price of oil in 1973–1974 was that OPEC, the Organization of Petroleum Exporting Countries, had gone too far; there would be a glut of oil on the world market, and the price of oil would soon decline again.[8] This reaction reflected the point of view that the price increase was due to a successful cartel that would eventually break down because of overproduction, and thus in the long run oil prices would tend to approach the costs of production in the main producing areas.[9] This point of view has been replaced to a considerable extent by the view that there still is a large potential for raising the price of oil because it is a finite resource and the costs of substitutes are generally well above the present price. Also, there is a growing awareness that there are discrepancies in the projected patterns of supply and demand for energy.[10] These may create shortages, particularly for oil, in the 1980s and could stimulate new sudden price increases.

With the depletion of conventional petroleum resources, the late-twentieth-century world has a relatively short period of time to organize the move from low-cost to high-cost energy.[11] The failure to organize this move properly will provoke new and more serious energy crises, severe setbacks for the world economy,

[8]Edward R. Fried, "World Market Trends and Bargaining Leverage," in Joseph A. Yager and Eleanor Steinberg (eds.), *Energy and U.S. Foreign Policy*, Ballinger, Cambridge, Mass., 1975, pp. 231–275.

[9]Ibid.

[10]*World Energy Outlook*, OECD, Paris, 1977, p. 8ff.

[11]F. Parra, Ramos, and A. Parra, *World Supplies of Primary Energy 1976–1980*, Energy Economics Information Service Ltd., Workingham, Berkshire, England, 1976, p. 13, and Mason Willrich, *Energy and World Politics*, The Free Press, New York, 1975, p. 27 ff.

and possibly international conflicts. Conventional oil production may peak around 1990 for physical reasons.[12] Not only are alternative sources in general much more costly, but they also take a long time to develop. Current investment in new sources of energy is largely insufficient to offset the anticipated decline in conventional oil production, let alone to cover incremental energy demand. There is thus a discrepancy between the technological horizon and the market horizon for energy that market forces alone are not able to resolve. Unless a strenuous effort is begun at once, both the industrialized and developing consuming countries are likely to experience a prolonged period at the end of this century during which energy is both expensive and scarce.

Most of the traditional oil-exporting countries are also in a difficult situation. Their economic and social development is a complex long-term process, and oil is their only real asset. The insufficient development of alternative sources of energy creates an increasing pressure upon the oil resources of these countries. They therefore risk entering the next century with depleted oil resources, financial assets eroded by inflation, and much larger populations. However, for some oil producers, increasing production will lead to financial surpluses, which they have not yet been able to invest successfully; this makes these surpluses less attractive. The decision facing a surplus member of OPEC can be seen as a portfolio decision: whether to invest in oil in the ground by holding back production or to produce above current economic requirements and to invest abroad.

Thus, given this overall situation, there is a case for a political solution to the problems of the world oil market.[13] The purpose of this study is to propose a negotiated international oil agreement that could help overcome the discrepancy between the techno-

---

[12]Carroll L. Wilson (ed.), *Energy: Global Prospects, 1985–2000*, Report of the Workshop on Alternative Energy Strategies, McGraw-Hill, New York, 1977, p. 17.

[13]Neil H. Jacoby, "Oil and the Future: Economic Consequences of the Oil Revolution," *The Journal of Energy and Economic Development*, Autumn 1975, pp. 45–54, and Mason Willrich and Melvin A. Conant, "The International Energy Agency: An Interpretation and Assessment," in *The American Journal of International Law*, April 1977, pp. 199–223.

logical capabilities and the market needs for energy, and at the same time help with the economic development of the oil-exporting countries. As a basis for this proposal, this study contains an analysis of the existing world oil market—both its economic and its political features—and a description of various ways it might evolve in the near future. At the outset, a brief sketch of the basic line of argument seems in order before turning to the analysis itself.

The point of departure will be an examination of the inherent instability of the world oil market, which results from the structure of the supply-and-demand relationship. This contributes to the high degree of politicization in the market. I shall then analyze the politics of both the OECD (Organization for Economic Cooperation and Development) and OPEC sides of the market. This discussion of their interests and potential for internal cooperation and conflict will be followed by an analysis of their interdependence.

Using this model of the political economy of the oil market, I shall look to future developments and the various ways oil politics might evolve. A solution to many of the future problems that are projected lies in the oil agreement I propose in the concluding chapter. It would regulate the market for the benefit of both producers and consumers, taking into consideration the current complexity of North-South relations.

The essentials of the international energy agreement proposed here can be reduced to a few elements. The most important one is a rational relationship between oil prices and oil supplies. There should be a gradual increase in the price of oil linked to the growth of consumption, and, up to a given level, OPEC should guarantee increased supplies. The negotiated price for oil should also be designed to reach the cost of alternative sources of energy when pressure on oil reserves reaches a certain point. This satisfies the need of both consumers and producers to stabilize the market and helps ensure an orderly transition to alternative sources of energy.

The second element involves encouraging oil producers to expand supplies while at the same time providing a method to offset the burden on the balance of payments of the oil consumers. This can be done by guaranteeing OPEC investments in the OECD area against inflation, currency depreciation, and nationalization. This not only will provide a means of absorbing balance-of-payments deficits and surpluses but will also give oil exporters assured sources of income and a direct interest in the economic health of the consumers.

The third element consists of giving the producers a direct interest in the energy consumption of the oil importers, combined with an intensified effort to develop alternative sources of energy. The oil producers should be given incentives to invest in downstream oil operations and in alternative sources of energy in the OECD area.

The fourth point, like the third, combines the OPEC concern with diversifying their sources of income and the desire of the oil consumers to make the producers more dependent on the OECD economies. This can be achieved by the OECD area offering technical and organizational assistance to the oil producers, along with greater access to OECD markets.

An international energy agreement that includes these elements could make the interdependence between the oil producers and oil consumers more balanced, help ensure the economic health of both OPEC and OECD countries, and reduce the political conflict potential of the world oil market.

As their vulnerability increases, the need of the industrialized consumers for an agreement regulating oil prices and oil supplies grows, and this weakens their bargaining position. It is therefore unlikely that the industrialized consumers will be able to obtain an agreement on oil without giving considerable attention to OPEC demands for linking an energy deal to assistance to the less developed countries (LDCs). This raises the question of whether an isolated energy agreement is politically possible or whether it could only be brought about as part of a comprehensive agreement, including economic development and commodities. In any case, an energy agreement that does not take at least the energy interests of the LDCs into consideration seems politically inconceivable. In the long run it could even be in the interest of the industrialized consumers to associate the LDCs with an energy deal. As consumers, the LDCs have interests that are increasingly similar to those of the industrialized consumers, and the LDCs might in the long run have a mediating influence between oil exporters and the OECD area.

It is, of course, difficult to discuss oil politics without taking into account the situation in the Middle East. The persistence of the Arab-Israeli conflict is likely to produce an increasingly unstable world oil market and make an international oil agreement difficult to achieve. The solution discussed in this study presupposes that the Arab-Israeli conflict will have been settled by the early 1980s, either through negotiation or a new war.

TWO

The International Politics of Oil

THE RESOURCE BASE

Many economic analyses of the energy situation tend to treat the supply side lightly.[1] In reality, the distinction between supplies from renewable and finite sources is of fundamental importance. It affects both replacement costs and the supply of substitutes and has particular significance for the study of the world oil market.

The production and consumption of a renewable resource can be seen as a self-sustaining circular process in which the resource base and the supply potential do not get eroded by continuous use. Examples of renewable resources are hydroelectric power and solar energy. By contrast, the production and consumption of a nonrenewable resource is a process of depletion, the resource base and the supply potential declining with use.[2] Typical examples are oil and natural gas. Historically, with increasing demand, prices have gone up, technology has improved, exploration has increased, and new reserves have been brought forth.[3]

[1] William H. Miernyk, "Regional Economic Consequences of High Energy Prices in the United States," *The Journal of Energy and Development*, Spring 1976, pp. 213–239.

[2] Nicholas Georgescu-Roegen, "Energy and Economic Myths," *Southern Economic Journal*, vol. 41, no. 3, January 1975, p. 367.

[3] Hendrik A. Houthakker, *The Price of World Oil*, The American Enterprise Institute for Public Policy, Washington, D.C., 1975, p. 11ff.

Although this pattern might continue for a long time to come, the basic fact is that nonrenewable resources exist in finite quantities, whatever the price.

An important element of this observation is that all energy resources exist in different categories of availability and cost. The transition from one level to the other is not always smooth and is often characterized by discontinuities, or even abrupt changes. This means that the transition is often not an automatic and continuous process, but is marked by difficult periods of adjustment, with gluts or shortages occurring, affecting economic growth and the distribution of income. In economic terms, the price elasticity of supply for energy seems to vary considerably over time. That is to say that the amount that the price must rise to bring forth one more unit of energy varies enormously depending on the available source. For example, in the extreme, at certain periods increasing supplies are available at falling prices, whereas at other times supplies can decrease in spite of rising prices. In political terms this means that the relations of strength and the control of the industry can change dramatically over time. Thus the exploitation of a nonrenewable resource can be seen as a historical process in which at least two major phases can be distinguished: In the first, exploitation moves into more easily accessible areas; then there is a shift as the pressure upon the easily accessible areas builds up, and in the second phase exploitation moves into less easily accessible areas.

In the first phase, a finite resource is exploited where it is most accessible, and the cost of increasing production declines during this period. Because more is available at lower costs in new places, newcomers can establish themselves relatively easily, and it can be an advantage not to be tied up in older and more costly areas of production. This stimulates competition, and real prices tend to fall. In addition, it is difficult for the producers to establish an effective collusive agreement if they do not have control of the markets.[4]

In the second phase, when exploitation moves into less accessible areas, the cost of expanding production tends to in-

[4]Jean-Marie Chevalier, *Le Nouvel Enjeu Pétrolier*, Calmann-Levy, Paris, 1973, p. 18ff, and Charles F. Doran, *Oil, Myth and Politics*, The Free Press, New York, 1977, p. 29.

crease. Since more is available only at higher costs, newcomers have more trouble establishing themselves, and it is advantageous to be producing in the older and less costly areas. Competition is thus impeded, and real prices tend to rise. Furthermore, it is relatively easy for producers to establish collusive agreements, even if they do not control the markets. An important part of this phenomenon is that the movement of real prices can be much more dramatic than the changes in the costs of production. In fact, there need not be any systematic relationship between the price and the costs of production.

In the first phase, when more is available at declining costs, the cost of substitutes is of little relevance, and the cost of production in the most accessible areas can be a useful point of reference, so that prices tend to decline toward this level.[5] In the second phase, when exploitation occurs in less accessible areas, the cost of replacement becomes more relevant, and the price approaches the cost of substitutes. To make the point succinctly: The historical shift from the first to the second phase of exploitation implies that the point of reference for pricing shifts from the cost of production to the cost of substitutes. However, the actual changes in price during these two phases depend upon a number of factors, such as the relationship between supply and demand and the degree of competition or collusion in the industry.[6] The basis for the political control of the industry also changes with the historical shift; in the first phase the control of the market is important, and in the second phase it is the control of supplies that matters.

The world's energy supplies are now running into such a period of discontinuous adjustment. The oil market had its historical shift in about 1970, when the discovery rate began to decline and consumption for the first time exceeded the expansion of new reserves through new discoveries. From 1950 to 1970 the average rate of discovery of new oil reserves in the world, excluding the U.S.S.R., Eastern Europe, and China, was about 2,500 million

[5]Chevalier, op. cit., p. 20, and M. A. Adelman, *The World Petroleum Market*, The Johns Hopkins University Press, Baltimore, 1972, p. 195.

[6]Paul Leo Eckbo, *The Future of World Oil*, Ballinger, Cambridge, Mass., 1976, p. 2.

metric tons, equivalent to 18 billion barrels, a year. Since 1970 the discovery rate has been lower, about 2,100 million metric tons or 15 billion barrels a year.[7]

Oil discoveries have an uneven geographical distribution. The first major oil provinces were in the United States and Russia, and later even more important oil provinces, with lower costs of production, were discovered in the Middle East. The new oil provinces are smaller and scattered around the LDCs, Arctic areas, the continental shelves of Western Europe and North America, and in remote parts of the U.S.S.R. They all have costs of production that are higher than in the Middle East. In the tropical areas, that is, in many LDCs, costs of production are relatively low and lead times are relatively short. In the industrialized countries of the Northern Hemisphere, costs of production from new oil-producing areas are high and lead times generally are long.

The world's total recoverable oil reserves are now estimated to be approximately 88 billion metric tons, or 650 billion barrels.[8] This corresponds to about 34 years of production at present rates. This figure may be revised downward. In recent years, reserve estimates for most parts of the world have been reduced, particularly in the Western Hemisphere and the Middle East.[9] Since 1970, North American oil production has been declining. In the contiguous United States the resource base is so eroded that production seems likely to decline regardless of the price of oil, and the essential question here is the rate of decline.[10] In coming years several of the traditional oil producers in the Middle East and elsewhere may reduce their oil production, because reserves are being depleted.[11] It is also possible that Soviet oil production may peak around or shortly after 1980. Oil production is now increasing in a few areas, such as Alaska, the North Sea, and

[7]Carroll L. Wilson (ed.), *Energy: Global Prospects 1985–2000*, Report of the Workshop on Alternative Energy Strategies, McGraw-Hill, New York, 1977, p. 119ff.

[8]*International Petroleum Encyclopedia 1977*, The Petroleum Publishing Co., Tulsa, Okla., 1977, pp. 303–306.

[9]*Oil and Gas Journal*, December 27, 1976.

[10]*Petroleum Intelligence Weekly*, January 3, 1977.

[11]*World Energy Outlook*, OECD, Paris, 1977, p. 8.

Mexico. It is reasonable to assume that past rates of discovery can only be maintained through an extensive exploration effort, combined with an extremely high success rate.[12] It is also reasonable to assume that future discoveries of oil will be unevenly distributed geographically and that in many cases the costs of production will be extremely high.

In the immediate future, incremental oil production will have to come both from new, less accessible areas, such as Alaska and the North Sea, and from some of the traditional areas of production, such as Saudi Arabia. In a long-term perspective, it can be assumed that a ceiling will be reached on Saudi production. This will make the search for oil in less accessible areas and the development of substitutes even more urgent. This implies that long-term production costs can be seen as rising, but at present it is impossible to define the border between the short term and the long term in this respect. Consequently, price developments are also uncertain. Since the historical shift to the second phase has already taken place, the cost of substitutes is a useful point of reference in the long term. In the short term, other factors, such as the relationship between supply and demand and the preferences of important producers, especially Saudi Arabia, may be decisive. Thus, within the framework of a long-term upward trend, short-term and medium-term preferences and values can be decisive. In a word, politics are important to the price of oil.

A brief look at the history of the oil market illustrates the shift from the first to the second phase quite clearly. Energy costs can reasonably be assumed to have been declining from 1859 to 1970, as less expensive oil gradually replaced other forms of energy. This process accelerated after 1945, when not only the real price but even the nominal price of oil declined.[13] The result was a dramatic increase in the consumption and production of petroleum. Oil and natural gas accounted for 5 percent of the world's energy consumption in 1900 and for 62 percent in 1970.[14] Oil

[12]Wilson (ed.), *Energy: Global Prospects*, p. 125ff.

[13]Chevalier, *Nouvel Enjeu*, p. 19.

[14]Joel Darmstadter et al., *Energy in the World Economy*, The Johns Hopkins University Press, Baltimore, 1971, p. 106ff.

alone accounted for 4 percent in 1900 and for 44 percent in 1970. If oil consumption and production had continued to grow at past rates from 1970 to 1980, the total output of oil during this decade would probably not have been much less than the total quantity of oil produced and consumed from 1859 to 1970.[15]

The exponential growth rates of consumption and production in the first phase prepared the ground for the historical shift. About 1960 it became evident to the international oil industry that traditional oil-producing areas, such as the Middle East, would not be able to meet demand in the long run. So interest developed in new, less accessible areas like Alaska and the North Sea, where the cost of production then exceeded the price of oil. With the price increases, oil production in these less accessible areas has now become economical. Oil production in newer and more remote areas, such as the Siberian continental shelf, the Beaufort Sea, and even possibly the Antarctic, will require even higher oil prices.

In the second phase of oil production, the major point of reference for the price of oil is the cost of substitutes. The estimated cost of alternative sources of energy has grown over the past years because of unforeseen technical and environmental problems. A useful point of reference is the cost of synthetic oil. It is now estimated to be twice to three times the present price of oil, that is, in the range of $25 to $40 per barrel.[16] This now seems to be considered the ceiling for the future price of oil.

In the long term it is difficult to argue that the world is confronted with a shortage of energy or even of fossil fuels or hydrocarbons. Oil reserves could multiply if nonconventional sources are developed, such as tar sands, shale oil, and heavy oil. Coal might also provide an extensive basis for synthetic oil. The technology for the exploitation of these resources is improving; the main limitations are costs and capital. From this perspective, the world energy problem has less to do with the

[15]Christopher Tugendhat and Adrian Hamilton, *Oil—The Biggest Business*, Eyre Methuen, London, 1975, p. 215.

[16]Hannes Porias, "Alternate Sources of Energy: Possibilities and Constraints," in Ragaei El Mallakh and Carl McGuire (eds.), *U.S. and World Energy Resources*, ICEED, Boulder, Colo., 1977, pp. 75–86.

finite limits of conventional oil resources than the replacement of conventional oil and how this should be organized. A major difficulty with alternatives to oil is that not only are their costs rising but lead times are increasing as well. In addition, political constraints on the development of alternatives are growing. Examples are found in public concern over the safety of nuclear reactors and in environmental restrictions over the production and use of coal in many industrial countries. The combined impact of these factors may be such that in the short term the price elasticity of energy supplies may be very low; that is, incremental energy supplies might not be increased no matter how much prices rise in the short run. This clearly compounds the uncertainty over prices and supplies.

The amount of conventional oil available will determine the rate at which alternatives must be developed to ensure a smooth transition to other forms of energy. The actual rate of depletion of the conventional oil reserves is dependent on the future relationship between the rate of reserve expansion and the rate of growth of consumption. Estimates of future additions to world oil reserves are, of course, most uncertain, because they depend upon new discoveries and improved recovery from existing wells. The Workshop on Alternative Energy Strategies makes the following evaluation of possible additions to world oil reserves.

TABLE 1
Estimated Annual Additions to World Oil Reserves, in Billions of Metric Tons

	High	Low
1975–2000	2.8	1.4
2000–2010	1.7	1.1
2010–2020	1.0	0.8
2020–2025	0.6	0.4

SOURCE: Carroll L. Wilson (ed.), *Energy: Global Prospects 1985–2000*, Report of the Workshop on Alternative Energy Strategies, McGraw-Hill, New York, 1977, p. 126.

15

With the high rate of reserve expansion, the gross addition to world oil reserves could be 70 billion tons between 1975 and 2000, almost doubling the 91 billion tons of oil reserves that were available by the end of 1975. With the low rate of reserve expansion, oil reserves would grow by 35 billion tons between 1975 and 2000, increasing reserves by only slightly more than a third. These estimates for reserve expansion should, however, be measured against estimates for oil consumption to get a clear picture of future world oil reserves.

Here, four rates of reserve expansion will be assumed:

1. A very high rate, 25 billion barrels or 3.5 billion tons a year
2. A high rate, 20 billion barrels or 2.8 billion tons a year
3. A medium rate, 15 billion barrels, or 2.1 billion tons a year
4. A low rate, 10 billion barrels, or 1.4 billion tons a year.

Only the most optimistic of combinations, those with a very high rate of reserve expansion combined with a low rate of annual demand growth, prevent oil reserves from being seriously eroded in this century. With all other combinations, available oil reserves will be diminished considerably, particularly in the period after 1990. Some combinations, which cannot be ruled out, even give a physical deficit of oil by 2000. This means that the relationship between reserves and production is likely to decline considerably in the 1980s and 1990s, making for a potential shortage of oil by 1990 or perhaps even earlier. This would probably stimulate oil producers to adopt more restrictive depletion policies and encourage consumers to compete more agressively for available oil supplies. Such developments would inevitably call into question the availability of oil in an open world market.

THE DEPENDENCE ON OIL

At present the OECD area is heavily dependent on other countries for its oil. Furthermore, it is by far the world's biggest consumer of oil. In 1975 the OECD countries accounted for 66

TABLE 2
State of World Oil Reserves at Different Rates of Reserve Expansion and Demand Growth, in Millions of Metric Tons

Projected Reserves at a Very High Rate of Expansion

Growth rate of demand (%)	1975	1980	1985	1990	1995	2000
2	91,000	93,300	95,800	92,900	89,900	84,700
3	91,000	92,800	92,100	88,500	81,600	70,700
4	91,000	92,300	90,200	83,700	72,100	54,500
5	91,000	91,800	88,100	78,500	61,400	34,700

Projected Reserves at a High Rate of Expansion

Growth rate of demand (%)	1975	1980	1985	1990	1995	2000
2	91,000	89,800	87,800	82,400	75,100	67,200
3	91,000	89,300	85,100	78,000	67,600	53,200
4	91,000	88,800	83,200	73,200	58,100	36,700
5	91,000	88,300	81,100	68,000	47,400	17,200

Projected Reserves at a Medium Rate of Expansion

Growth rate of demand (%)	1975	1980	1985	1990	1995	2000
2	91,000	86,300	81,800	71,900	61,900	49,800
3	91,000	85,800	78,100	67,500	53,600	35,700
4	91,000	85,300	76,200	62,700	44,100	19,200
5	91,000	84,800	74,100	57,500	33,400	(−300)

TABLE 2 (Continued)
State of World Oil Reserves at Different Rates of Reserve Expansion and Demand Growth, in Millions of Metric Tons

	Projected Reserves at a Low Rate of Expansion					
Growth rate of demand (%)	*1975*	*1980*	*1985*	*1990*	*1995*	*2000*
2	91,000	82,800	74,800	61,400	47,900	32,200
3	91,000	82,300	71,100	57,000	39,600	18,200
4	91,000	81,800	69,200	52,200	30,100	1,700
5	91,000	81,300	67,100	47,000	19,400	(−17,400)*

*Negative values indicate the extent to which demand will have exceeded reserves.

SOURCE: Author's own calculations.

percent of world oil consumption, but only 25 percent of production. Imports into the OECD area represented close to 45 percent of all oil produced in the world. The OECD area has 10 percent of the world's proven oil reserves, corresponding to 17 years of production and to 5 years of consumption at current rates. By contrast, in the rest of the world oil reserves correspond to 38 years of production and 87 years of consumption.[17]

This has triggered an intense debate about the nature of Western dependence on imported oil. There are profound disagreements over the outlook for the medium term, the period from 5 to 15 or 20 years ahead. First, there is disagreement about the impact of price hikes on the general level of energy consumption and, in particular, on the relationship between the rate of economic growth and the growth of energy consumption. Second, there is disagreement on the substitution of expensive imported oil for other forms of energy. The controversy can be summed up by two opposing arguments.

[17]*International Petroleum Encyclopedia 1977*, p. 303.

One view is that there are still large quantities of inexpensive energy available. The increase in the oil price will first lead to a reduced growth of demand and then to an increasing output of oil, from the discovery of new fields and from improved recovery in existing oil fields. The higher prices will also accelerate the development of alternative sources of energy whose costs will fall. There may even be a glut of energy, and particularly of oil. Excess capacity will cause OPEC to break down because of disagreements on the distribution of production and income.[18] Eventually, prices will fall and Western dependence on imported oil will be seen as transitory.[19]

The second argument points out that oil is a scarce resource. Other forms of energy either are technically undeveloped (solar energy, fusion), are very expensive (synthetic oil), or cause environmental and safety problems (coal, nuclear fission). The demand for energy does not seem to be too sensitive to price changes, at least within the known limits. Historically, there is a fairly close relationship between the level of energy consumption and the level of economic activity, even if there are important variations by country and over time.[20] Because probabilities of improving the discovery rate for oil, or of substantially improving the recovery rate from existing fields, are low, and because of the large costs and lead times for other forms of energy, there is unlikely to be a glut of oil, or of energy. On the contrary, there may be a shortage of energy until alternatives are sufficiently developed, and chances are that the price of oil will increase further.[21] The dependence of the Western industrial countries on imported oil is likely to continue for a considerable period of time, at even higher prices.[22]

---

[18]Eckbo, *Future of World Oil*, p. 269ff.

[19]Edward R. Fried, "World Market Trends and Bargaining Leverage," in Joseph A. Yager and Eleanor Steinberg (eds.), *Energy and U.S. Foreign Policy*, Ballinger, Cambridge, Mass., 1975, p. 269ff.

[20]Darmstadter et al., *Energy in the World Economy*, p. 27ff.

[21]Parra, Ramos, and Parra, *World Supplies of Primary Energy 1976–1980*, Energy Economics Information Service Ltd., Wokingham, Berkshire, England, 1976, p. 13.

[22]*World Energy Outlook*, p. 9.

The first argument fits well with neoclassical economics, and the second reflects certain neo-Malthusian assumptions. Differences in relative endowments and in ideological propensities may explain why neoclassical analyses and prescriptions are more accepted in North America than in Western Europe.

The choice of argument also has a practical importance in the choice of policies and solutions. If the first argument is right, the high relative price of energy and dependence on imported oil will last a short time, constituting a temporary disturbance for the OECD economies. This disturbance can be most properly countered by short-term measures designed to neutralize the effects of the high price of oil.[23] If, however, the second argument is more valid, the high relative price of energy and the dependence upon oil imports will last a long time and will reflect a structural change in the environment and the fundamental working conditions of the OECD economies. In this case the appropriate response is by measures designed to adjust the OECD economies to the high price of oil. Structural changes in the OECD economies, and possibly in the international financial system, may be required. This may also imply an international negotiated settlement to regulate the world oil market.

Choosing between the two arguments is risky because the implementation of the wrong policy could be costly. Applying long-term measures to a temporary phenomenon does more harm than good by keeping the relative price of oil at an artificially high level, thus reducing the possible rate of economic growth in Western countries and LDCs. Applying short-term measures to a structural problem may have dangerous consequences because it would keep the Western economies and possibly the LDCs in an artificially comfortable environment in which contradictions would build to create an extremely serious future energy shock.

The disagreement between the two positions essentially comes down to differences over the long-term price elasticities of demand and supply for energy. The "neoclassical" argument as-

[23]Eric Rhenman, *Organisationsproblem och långsiktsplanering*, Bonniers, Stockholm, 1975, p. 26f.

sumes that both are high, which means that both supply and demand are responsive to changes in price. This is believed because it is rational economic behavior to consume less and to produce more energy at higher prices. The "neo-Malthusian" argument assumes that both are low, or in other words that demand and supply are not very responsive to changes in price. This assumption is based on the observation that consumers are largely indifferent to energy prices—giving a high priority to energy-intensive habits of consumption. In addition, the physical and technical possibilities for increasing energy output are limited. So the major points of discord relate to consumer behavior in the face of rising prices, the probability of finding new oil fields or improving recovery from existing ones, and the prospects for alternative energy sources.

The two arguments are not mutually exclusive in a dynamic perspective. It can reasonably be argued that in a medium-term perspective the neo-Malthusian position provides the best image of reality and that the neoclassical pattern will only assert itself over the long run. As a result, policy makers should not assume that the forces of the market will cause OPEC to break down and oil prices to fall again as neoclassical analysis suggests.[24]

Two observations support this dynamic view of the neo-Malthusian and neoclassical positions. First, the high price of oil set by OPEC is essentially a function of external conditions, namely the historical shift of the exploitation to less accessible areas. By doing away with OPEC, the process might be delayed somewhat but not fundamentally changed. Eventually there will be a transition to other forms of energy and the market will adjust to this in the long run.

Second, the direct substitution of other forms of energy for oil is difficult for many reasons. Oil is in many ways an ideal primary energy source. It is relatively clean, is transportable, and allows consumers considerable flexibility. These technical qualities have fostered specific patterns of consumption and have

---

[24]Ragaei El Mallakh et al., *Implications of Regional Development in the Middle East for U.S. Trade, Capital Flows and Balance of Payments*, ICEED, Boulder, Col., 1977, p. 9.

increased the use of energy in modern economies. This can be seen by the rapid increase in Western oil consumption in the postwar period, which is not due to the replacement of coal by oil as much as it is the result of new patterns of oil consumption. The use of the automobile, petrochemical products, and oil-based heating systems are the best examples of these new patterns. The direct substitution of oil by other primary sources of energy is possible in only some cases. Coal can replace oil for heating, and both coal and nuclear power can be used to produce electricity, but there are no primary forms of energy that can substitute for most transportation and petrochemical uses of oil. A good substitute for oil must have essentially the same characteristics of flexibility, transportability, and so forth. This suggests synthetic oil, which is extremely expensive and much less efficient than primary sources of energy. The widespread substitution of oil by coal and nuclear, geothermal, solar, and other forms of energy, implies extensive changes in production processes, patterns of consumption, and methods of transportation. This means large investments and high capital costs, and it would only be economical at very high oil prices.

Given these realities, it is no wonder that it is hard to find good substitutes for oil, and this explains the low price elasticity of demand generally observed.[25] Oil, quite simply, is a necessary input in many modern production and consumption processes. The importance of oil is not reflected by its small part of the gross national product,[26] but it nonetheless functions as a catalyst without which other inputs would be much less effective.[27] It should be recalled that substantial conservation efforts require a long time. For example, reducing by half the amount of energy

[25]Neil H. Jacoby, "Oil and the Future: Economic Consequences of the Oil Revolution," *The Journal of Energy and Economic Development*, Autumn 1975, p. 15.

[26]Chauncey Starr, *Energy Planning—A Nation at Risk*, Electric Power Research Institute, Palo Alto, Calif., 1977, p. 2.

[27]William W. Hogan and Alan S. Manne, *Energy-Economy Interactions: The Fable of the Elephant and the Rabbit*, Energy Modeling Forum, Stanford University, Stanford, 1977, p. B-2ff.

needed for producing ammonia has taken 60 years.[28] The replacement of one form of energy by another as the dominant one is a long historical process too. It took 70 years for oil to move from providing 4 percent to providing 44 percent of the world's energy. In the medium term there are obviously limits to the ability of substitutes or conservation to replace oil. This unavoidable dependence on oil, which necessarily follows, makes the availability of oil a critical political question for all countries and places us squarely in the neo-Malthusian dilemma, at least for the short run.

THE POLITICIZATION OF OIL

The critical position of oil in the world's energy balance and the uneven distribution of reserves give oil a lot of economic, strategic, and political importance. The oil price and the control of supplies are thus potentially conflicting political issues. In addition, oil becomes linked directly or indirectly to other issues. Since most countries are net importers of oil and rely heavily on this oil for total energy supplies, the price of oil and its control can have direct implications for their freedom of action in economic and foreign policy. Oil is thus linked to such matters as the rate of economic growth, the level of employment, the rate of inflation, trade policy, and general foreign policy orientation. It is also linked to the political cohesion of the world's political blocs and to the development of the LDCs.[28a] Consequently, matters relating to oil have a high priority in the industrial, economic, trade, and foreign policies of both oil-importing and oil-exporting nations, whether they are developed or developing economies.

The political importance of oil means that changes in the international oil market can have consequences for the international distribution of power. The best example is provided by the Arab oil-exporting countries, which in the 1970s have dramatically improved their ability to pursue foreign-policy goals. Another example is the United States, which by organizing the

[28]Porias, "Alternate Sources of Energy," p. 84.
[28a]Willrich, *Energy and World Politics*, p. 180ff.

industrial oil-consuming countries in the International Energy Agency (IEA) hoped to offset any loss to its own position of leadership in the Western world inflicted by the new conditions in the oil market.[29] Thus the structure and organization of the world oil market not only serve purposes of rationality and efficiency but are also in part mechanisms of political control.[30]

For both OECD and OPEC countries there can be internal conflicts over priorities related to oil and other social, economic, trade, and foreign policy goals. This can create difficult choices. Restricting or expanding oil imports can directly affect domestic political goals in OECD countries. Conversely, the world's oil exporters are concerned about the rate of depletion of their finite resources. Several of the traditional oil-exporting countries might not be able to sustain past or present levels of production, and others may prefer to keep the major national asset in the ground, deferring income to future generations.[31] The way they exploit their oil plays a decisive role in determining how and when they influence the market. Securing supplies and controlling demand are of primary importance to the governments of consumer countries, and the management of reserves is critical to the governments of the oil-exporting countries if they seek to maximize their income or promote other political goals. As a result the production, distribution, and consumption of oil are to a large extent, and increasingly, subject to government intervention and regulation.

The international oil industry has an oligopolistic structure because it is dominated by a limited number of private international and state-owned oil companies. Their long-term interests and policies often determine their short-term behavior.[32] In addition, the oil companies and the governments normally have close relations of consultation and cooperation. The major agents on the international oil market are thus a limited number of

[29]Martin Saeter, "Oljen og de politiske samarbeidsformer," *Internasjonal Politikk*, no. 2B, 1975, pp. 397–421.

[30]Ibid., p. 397.

[31]*World Energy Outlook*, p. 8ff.

[32]Anthony Sampson, *The Seven Sisters*, Hodder and Stoughton, London, 1975, p. 3ff.

companies and governments, often with clearly defined long-term interests. Consequently, the patterns of oil production, distribution, and consumption are less the result of market forces than is the case with many other commodities. Instead, political intervention and long-term considerations have a greater impact, and the nation-state is a more important agent.[33]

The oil crisis of 1973–1974 demonstrated the complexity and wider political significance of the oil market. The Western-dominated system of relations between OECD and OPEC countries, and between the OECD area and the Third World, was shaken. The Arab oil embargo was explicitly linked with the conflict in the Middle East and with the position of the OECD countries toward Israel. This in turn directly affected relations between the United States and Western Europe. The cooperative policy of the European Community (EC) toward the Arab and the Mediterranean countries was in conflict with the Atlanticist policy of consolidation of the United States.

The foundation of the IEA affected the EC plans for a common energy policy and the process of economic and political integration in the EC. The IEA can be seen not only as an attempt to solve the energy problems of the OECD countries on a common basis but also as an attempt to mold the OECD countries into an institutional framework controlled by the United States.[34] One purpose was to prevent extensive bilateral deals between the other OECD countries and the oil exporters, because this could dilute the cohesion of the OECD area and reduce the political influence of the United States. Another purpose was to preserve the position of United States–based multinational oil companies in supplying OECD countries with oil. An extensive network of bilateral deals between the other OECD countries and the oil producers would probably affect the structure of international oil trade, reducing the role of the multinationals and benefiting the national oil companies of the producing and consuming countries. The solution for the United States, to defend its interests, was to attempt to speak for all the industrialized

[33]Tugendhat and Hamilton, *Oil*, p. 250ff.
[34]Ibid., p. 409.

consumers. This in part explains why the United States, which is much less dependent on imported oil than Western Europe and Japan, has, at least verbally, taken a much more aggressive attitude toward OPEC.

Another dimension in the split within the OECD over how to respond to OPEC concerns Israel. The United States has consistently supported Israel in the Middle East and was reluctant to back down simply because of the oil embargo. The other OECD countries tended to compromise with the Arab oil producers more easily. France's support for the Palestinians and its bilateral arrangements with Arab countries are probably the best example of this type of response. It is clear that just as the Atlanticist policy was in the best interest of the United States, the cooperative policy of the EC was likewise a response of self-interest. The relatively low dependence of the United States on foreign oil allowed it to continue its support of Israel while the Europeans, with a greater dependence on Middle East oil, preferred to avoid confrontations with the Arab oil producers over Israel. The Europeans also knew that the United States was not going to abandon Israel. In this context they could use their softer stance in an attempt to placate the Arab oil producers and make bilateral deals with them to help provide for their immediate energy needs. In the future this split could have dramatic consequences for the cohesion of the OECD area, especially in the event of a new war or prolonged tension in the Middle East that threatens the security of supplies. If the United States felt that it had to intervene directly this might even lead to several European countries leaving the North Atlantic Treaty Organization (NATO).

Similarly, splits could develop over policies toward the Third World and the structure of a new international economic order, especially if OPEC countries directly linked negotiations over North-South relations to oil prices and supplies. Some European governments—that of France, for example—believe that a more generous attitude toward Third World demands could have a beneficial impact on the security of their oil supplies. The Scandinavian countries and the Netherlands support a more generous attitude toward Third World demands, for ethical and ideological reasons. Thus a split could develop within the OECD area between generous and less generous friends of the Third World, with the first perhaps hoping to have preferential treatment from OPEC and preferred access to LDC markets and resources.

It is significant that the links to other important economic and political problems have increasingly been made explicit, whereas only a few years ago they were only implicit. The United States has on several occasions stressed the link between oil and security in relations with Western Europe. In order to make Western Europe accept common institutions for energy policy, the United States used security policy and the presence of American troops in West Germany as a means of pressure.[35] Another recent example of the open linkage of issues was at the OPEC meeting in Doha in December 1976, where Saudi Arabia explicitly linked the price of oil to Middle East peace developments and to the results of the Conference on International Economic Cooperation (CIEC) held in Paris in 1976–1977.[36]

OIL REGIMES

The dramatic character of the oil crisis of 1973–1974 overshadowed some of the long-term economic problems with the demand and supply of energy. The crisis produced in a few months important economic and political changes in the world oil market. These changes were a kind of "oil revolution,"[37] a transition from the First Oil Regime to the Second Oil Regime.

The First Oil Regime was characterized by an integrated pattern of organization, based in the major consuming countries, and a low price for oil. During the First Oil Regime the center of the world's oil production gradually shifted from North America to the Middle East. Decreasing exploitation costs and the political dominance of important producing areas by major consuming countries made this shift possible. The industrialized

[35]Ibid., p. 402.
[36]Louis Turner, "Oil in the North-South Dialogue," *The World Today*, February 1977, p. 57.
[37]Tugendhat and Hamilton, *Oil*, p. 179ff.

consuming countries became increasingly dependent on a limited number of developing countries for their oil. Within the First Oil Regime the basis of the regime's existence was eventually eroded because power shifted to countries whose interests the regime did not serve. The combination of rapidly growing demand and rising exploitation costs (shown by the investments in areas like Alaska and the North Sea) proved fatal to the First Oil Regime because it opened the way for a price increase and institutional change through OPEC control of production. This loss of the economic and political basis of the regime explains the abruptness of the transition, once the catalyst of the Middle East dispute set it off.

The Second Oil regime is characterized by a fragmented pattern of organization and a much higher price for oil. The center of world oil production is the Middle East, but for physical and political reasons it is not clear if growing demand can be met by supplies from this area. Efforts are being made to find and produce oil in new oil provinces, but it is unlikely that oil from these new areas will be able to keep pace with demand for long. Thus the need to develop alternative sources of energy will become more and more urgent.[38] The Second Oil Regime, like the first, erodes the basis of its own existence through its inability—for physical and political reasons—to guarantee sufficient supplies of oil.

Assuming that in the short run the industrial consuming countries fail to achieve much greater energy self-sufficiency, a sudden collapse of the Second Oil Regime can be anticipated. One of the following three options seems probable:

- The oil-producing countries will supply the quantities of oil demanded on the world market at gradually higher prices so that alternative sources of energy eventually become more competitive.
- The oil-producing countries will supply the quantities of oil demanded, but at sharply increasing prices, provoking a new recession in the OECD area.

[38]Ibid., p. 380.

- The oil-producing countries will not supply the quantities of oil demanded on the world market but will use oil as a means to achieve other political goals by rationing exports and selecting favorite clients.

It is clearly in the interest of the OECD countries to avoid the last two possibilities, even if this means the end of the Second Oil Regime through an accelerated development of alternative sources of energy or through direct military intervention in the oil-producing countries. It is in the interest of both producers and consumers to have a gradual transition from the Second Oil Regime to a third one; indeed, such a transition appears to be inevitable. A historical trend for incremental energy supplies can be assumed, moving from cheap conventional oil, by way of more expensive oil, to extremely expensive synthetic oil. This trend can also provide the basis for a transition from one oil regime to another. These three regimes are schematically presented in Table 3.

Regardless of how we arrive at the Third Oil Regime, it is clear that the energy problems that caused the oil revolution have not vanished, and the Second Oil Regime has not yet been tested in a situation of rapidly increasing demand. Therefore, the political and physical limits to the energy supplies of the OECD countries

TABLE 3
The Three Oil Regimes

	Oil Regime		
	First	*Second*	*Third*
Incremental resource base	Conventional oil	Alternative oil	Synthetic oil
Marginal costs	Falling	Rising	Rising
Price	Low	High	Very High
Structure	Integrated	Disintegrated	?
Control	Consumers	Producers	?

29

could be tested again.[39] In practical terms this means that their bilateral relations with OPEC countries and the development of alternative sources of energy are of primary concern to the OECD area. The crucial question for this study is whether the OECD countries and OPEC are better off with or without a negotiated settlement. To resolve this question we need a clearer picture of OECD-area needs and the likely OPEC response to them.

TABLE 4
Distribution of World Oil Reserves and Oil Production in 1975, in Millions of Metric Tons

	Reserves	*Production*	*Ratio of reserves to production*
North America	5,656	489	11.5
Western Europe	3,531	27	130.7
Middle East	55,247	977	56.5
Latin America	5,551	219	25.5
Asia	2,879	110	26.1
Africa	9,343	249	37.5
Communist nations	15,239	592	25.7

SOURCE: *International Petroleum Encyclopedia 1976,* The Petroleum Publishing Co., Tulsa, Okla., 1976, pp. 12–13, and *Basic Petroleum Data Book,* American Petroleum Institute, Washington, D.C., 1975, section IV, table 1.

[39]*World Energy Outlook*, p. 8ff.

TABLE 5
World Petroleum Trade Balances for 1976, in Millions of Metric Tons

Region	Petroleum production	Petroleum consumption	Trade balance
North America	489.1	908.3	− 419
Western Europe	46.5	691.1	− 645
Australia, New Zealand	22.8	37.7	−15
Japan	0.5	254.5	− 254
OECD area total	558.9	1,891.6	− 1,333
Middle East	1,137.8	88.5	+ 1,050
Asia*	115.5	121.2	−6
Africa	291.0	54.2	+ 237
Latin America	228.4	189.1	+39
China	91.0	66.0	+27
U.S.S.R. + Eastern Europe	527.3	460.0	+67
World excluding OECD	2,391.0	979.0	+ 1,414
World including OECD	2,949.9	2,870.5	

*Excluding Japan, Australia, and New Zealand.
SOURCE: *International Petroleum Encyclopedia, 1977*, The Petroleum Publishing Co., Tulsa, Okla., 1977, pp. 392–393, 277–279.

TABLE 6

World Energy and Oil Consumption, 1900–1974

Year	World energy consumption (millions of metric tons of oil or the equivalent)	Yearly growth rate	Oil consumption (millions of metric tons)	Yearly growth rate	Cumulative Oil consumption by period (millions of metric tons)	Cumulative oil consumption (millions of metric tons)	Period as a percentage of total
1900	532		21		451	451	1.3
1913	945	4.52	57	7.98	1,645	2,096	4.6
1929	1,190	1.45	179	7.41	1,592	3,688	4.4
1937	1,260	0.72	227	3.01	4,070	7,758	11.4
1950	1,750	2.56	438	5.19	4,674	12,432	13.0
1958	2,632	5.23	816	8.09	9,917	22,349	27.7
1967	3,976	4.69	1,551	7.35	13,472	35,821	37.6
1974	5,579	4.96	2,511	7.12			

SOURCE: John Chesshire and Keith Pavitt, *Social and Technological Alternatives for the Future—Energy*, Science Policy Research Unit, University of Sussex, Brighton, England, 1977, p. 9.

TABLE 7
World Petroleum Consumption and Production 1950–1975,
in Millions of Metric Tons

	Consumption				Production		
Year	World	WECC*	OECD		World	WECC*	OECD
1950	478	436	368		514	476	274
1960	1,051	907	753		1,036	875	388
1970	2,281	1,948	1,608		2,216	1,831	562
1973	2,789	2,355	1,949		2,740	2,284	578
1975	2,742	2,239	1,804		2,622	2,039	527

*World excluding the Communist countries (U.S.S.R., Eastern Europe, and China).
SOURCE: *International Petroleum Encyclopedia, 1977*, pp. 392–393, 277–279.

TABLE 8
World Oil Reserves and Oil Production 1950–1975,
in Millions of Metric Tons

Year	World oil reserves	World oil production	Ratio of reserves to production
1950	10,458	520	20.1
1960	39,676	1,049	37.8
1970	72,576	2,295	31.6
1973	90,864	2,786	32.6
1975	97,458	2,664	36.5

SOURCE: *International Petroleum Encyclopedia, 1977*, pp. 12–13, and *Basic Petroleum Data Book*, section IV, table 1.

TABLE 9
Rates of Growth of Oil Consumption

	World	WECC*	OECD
1950–1960	8.2%	7.6%	7.4%
1960–1970	8.1	7.9	7.9
1970–1973	6.9	6.5	6.6
1950–1973	8.0	7.6	7.5
1960–1973	7.8	7.2	7.6

*World excluding the Communist countries (U.S.S.R., Eastern Europe, and China).

SOURCE: Chesshire and Pavitt, *Social and Technological Alternatives for the Future—Energy*, p. 8.

TABLE 10
Cumulative Volume of Oil Consumed,
in Millions of Metric Tons

	World	WECC*	OECD
1950–1959	6,989.2	6,198.8	5,174.5
1960–1969	15,299.3	13,078.5	10,859.6
1970–1973	10,112.1	8,583.0	7,099.0
1950–1973	32,400.6	27,860.3	23,133.1
1960–1973	25,411.4	21,661.5	17.958.6
1970–1980†	33,199.7	28,079.0	23,189.8

*World excluding the Communist countries (U.S.S.R., Eastern Europe, and China).

†Hypothetical at growth rates of period 1960–1969.

SOURCE: Chesshire and Pavitt, *Social and Technological Alternatives for the Future—Energy*, p. 8.

THREE

OECD Demand

Under the First Oil Regime low prices and increasing supplies of oil gave most OECD governments inexpensive and assured energy supplies without disturbing important economic and foreign policy goals. Cheap foreign oil covered their incremental energy demands and substituted for more expensive domestic sources of energy, without weighing heavily on their balance of payments. Their control of the oil industry implied that increasing dependence on oil imports was not associated with political dependence on the oil-exporting countries. For example, as recently as during the Six-Day War in 1967, oil supplies from the Middle East were generally uninterrupted. Thus the First Oil Regime allowed OECD governments to ignore the problem of potential oil shortages.

In the post-1974 period, under the Second Oil Regime, continued economic growth in the OECD countries could be less compatible with other important goals of economic and foreign policy. High oil prices could have a negative effect on the balance of payments and, indirectly at least, stimulate inflation. Increasing dependence on oil imports could also limit freedom of action in foreign policy. At the same time, though, high levels of unemployment and stagnant real incomes in many OECD countries make it difficult for governments to pursue a policy of economic austerity over a long period. Many OECD governments, if not most, may thus have to sacrifice internal or external policy goals to cope with their energy problems.

The crucial question is: To what extent does the oil revolution actually mark a historic break in energy demand trends? In 1974 and 1975 there was an absolute decline in the energy consumption of the OECD area, caused almost entirely by the drop in oil consumption. This decline was due less to the price increase than to the economic recession and to the exceptionally mild winters. Since the end of 1975 energy consumption, including the use of oil, has increased again as economic growth has picked up. Future growth rates will depend on national and international market forces, on the domestic political pressures put on governments, and on their freedom of action in the relationship between economic policy and energy policy.

THE FACTORS OF DEMAND

The important economic factors behind oil imports are the rate of economic growth, the domestic output of energy, and the energy coefficient (the relationship between the economic growth rate and the growth of energy consumption). In terms of economic analysis, oil imports are the residual, or what is left over, of the economy's energy requirements, after domestic energy production has been absorbed.

The important political factors that influence oil imports are the goals of national economic and energy policies and the impact of business and public interest groups on these policies. Oil imports can be seen as unmet national energy needs, which are one element in the balancing of a nation's economic and energy goals in such a way that excessive domestic opposition is avoided.

These two approaches to the analysis of oil imports are complementary. One stresses the importance of economic processes, the other the importance of political decisions. Both underline the fact that oil imports represent that part of essential national energy needs that cannot be satisfied domestically. This implies that relatively small economic or political changes can have quite important effects on the level of oil imports. Let us now turn to the specific economic processes that influence demand for oil in the OECD area.

The energy coefficient is critical, and the OECD economies can reasonably be expected to have a lower one in the post-1974 period than before 1973. In other words, pre-1973 economic growth rates can now be achieved with a smaller growth rate of energy consumption. Historically, there is a close relationship between the rate of economic growth and the growth of energy consumption, but there are important differences over time and between countries.[1] There has in some cases been a tendency for the energy coefficient to fall with greater industrial maturity.[2] Today the consumption of energy in relation to gross national product is twice as high in the United States as in West Germany or France.[3] Geographical size is not the only explanation for this. Both the price of energy and the content of economic growth are important factors here. Thus past relationships between economic growth and the growth of energy consumption cannot be uncritically projected into the Second Oil Regime. In economic terms, it is rational behavior to use less energy at high prices, and politically many governments are encouraging energy conservation.

There are also arguments against the trend toward declining energy coefficients. Consumers appear to give a high priority to habits of consumption that are energy-intensive, particularly with rising living standards.[4] Also, there is a widespread consumer indifference to rising costs and the insecurity of energy supply.[5] Industry is now reluctant to conserve energy by changing production processes and by new investments in energy-saving

[1]Darmstadter et al., *Energy in the World Economy*, The Johns Hopkins University Press, Baltimore, 1971, p. 37.

[2]John Chesshire and Keith Pavitt, *Social and Technological Alternatives for the Future—Energy*, Science Policy Research Unit, University of Sussex, Brighton, England, 1977, p. 6ff.

[3]Joel Darmstadter, Joy Dunkerley, and Jack Alterman, *How Industrial Societies Use Energy: A Comparative Analysis*, Resources for the Future, Washington, D.C., 1977, p. 21ff.

[4]D. K. Verleger and D. P. Sheehan, "A Study of the Demand for Gasoline," in Dale W. Jorgensen (ed.), *Econometric Studies of U.S. Energy Policy*, North Holland Publishing Company, Amsterdam, 1976, pp. 179–234.

[5]Anthony Sampson, *The Seven Sisters*, Hodder and Stoughton, London, 1975, p. 29.

equipment because of uncertainty over future energy prices and because the cost of energy conservation could be very high. In addition, many firms simply offset higher energy costs by higher prices. Finally, expenses for energy still make up only a small proportion of users' budgets, so price increases in many cases can continue to be absorbed.

Little is known about long-term energy coefficients at high income levels after a sharp price increase such as the one experienced in 1973. The OECD assumes an average energy coefficient of 0.84 over the period 1974–1985, at constant oil prices.[6] This is a decline of 24 percent in the energy coefficient of the pre-1973 period, when real oil prices were about one-third of what they are today.

We can reasonably suppose that the growth of domestic energy production in the OECD countries will be higher in the post-1974 period than before 1973. The question again is how much, and again little is known about the long-term price elasticity of energy supply in advanced industrial countries after a sharp price increase. In economic terms, it is rational to produce more energy at higher prices. Politically, many governments actively encourage domestic energy production to offset balance-of-payments deficits and reduce supply insecurity. In practice, however, the responsiveness of OECD energy production to the price increase seems to be modest.[7] There is the problem of rising costs, which applies to all forms of energy produced in the OECD area. Politically, there is increasing environmental concern, and uncertainty in the private sector. This is reflected in a general reluctance to channel large funds into the energy sector and in resistance to particular forms of energy production, such as nuclear power, coal, and offshore petroleum. Thus it is questionable to what extent market forces alone will be able to bring about a significant growth in energy production.

The potential production of oil in the OECD area will be influenced by new discoveries and by the energy policy of a few countries. There are definite possibilities of finding additional

[6]*World Energy Outlook*, OECD, Paris, 1977, p. 27.
[7]Ibid., p. 8ff.

reserves of petroleum in the OECD area. These possibilities are greatest in the new oil provinces—on the continental shelves of the United States, Canada, Great Britain, and Norway, as well as in the Canadian Arctic and Alaska. These regions are typical provinces of "alternative" oil, with high costs and lead times of 4 to 10 years.[8] Before 1985 these new fields are not likely to contribute substantially to the oil production of the OECD area. Another problem is the potential effect of changes in energy policy in key countries, particularly in the United States. For Americans the major question is the potential impact of deregulation of oil and gas prices on reserve estimates and supplies. In Canada the possibility of stopping the decline in oil production before 1985 seems remote. In Norway the question is the potential effect of a less restrictive production policy. In Great Britain the uncertainty concerns reduced output stemming from a possible depletion policy, once self-sufficiency is achieved.

In the period 1960–1973 the energy production of the OECD area expanded at an average yearly rate of 2.5 percent. For the period 1974–1985 the OECD assumes an average yearly growth in energy production of 3.5 percent, implying a strenuous effort in the form of political resolution and capital mobilization. If a much stronger effort is made, the OECD assumes that the growth rate could be 4.5 percent.[9]

REGIONAL CONTRASTS

The prospects for energy production are different in the various OECD regions, depending on local physical, economic, and political conditions. In the 1960s domestic energy production had a high yearly growth rate in North America, was stagnant in Western Europe, and was negative in Japan. By contrast in the 1970s domestic energy output has had a very low growth rate in North America, an exceptionally high growth rate in Western Europe,

[8]Christopher Tugendhat and Adrian Hamilton, *Oil—The Biggest Business*, Eyre Methuen, London, 1975, p. 332ff.
[9]*World Energy Outlook*, pp. 8–9.

TABLE 11
Average Yearly Growth Rates of Energy Production, 1960–1980, in Percentages

Region	1960–1970	1970–1980	1974–1980	1960–1980
North America	4.5	1.0	1.2	2.7
Western Europe	—	6.3	8.6	3.1
Japan	− 2.2	4.5	5.5	1.1

SOURCES: *Oil—The Present Situation and Future Prospects*, OECD, Paris, 1973, and *World Energy Outlook 1977*, OECD, Paris, 1977.

and a high growth rate in Japan. During the 1960s the differences in endowment of natural resources were decisive: North America had economically recoverable reserves of coal, oil, and gas, whereas Western Europe and Japan had declining coal industries and little or no oil and gas. In addition, the regulation of oil imports kept oil prices higher in North America than on the world market, which stimulated production. During the 1970s differences in energy policies seem more decisive: In Western Europe and Japan resolute energy policies are being implemented, while in North America, particularly in the United States, energy policies are more difficult to implement. Also, in the United States price regulation keeps oil prices lower than on the world market, which tends to discourage production. In addition, the economically recoverable reserves of oil and gas, at least in the traditional areas of production in North America, are declining; Western Europe, on the other hand, is tapping its new offshore reserves.

These differences in the growth of energy production give remarkable contrasts in the growth of oil imports (see Table 12). The contrasts are particularly striking over the years immediately following the oil crisis—the period 1974–1980. This period is characterized by an almost stagnant output of energy but rapidly increasing oil imports in North America and rapidly growing energy production with a slight drop in oil imports for Western Europe. To a certain extent these differences are explained by factors that are independent of the oil crisis, including investment

TABLE 12
Average Yearly Growth Rates of Oil Imports, 1960–1980, in Percentages

Region	1960–1970	1970–1980	1974–1980	1960–1980
North America	5.2	11.6	9.9	8.4
Western Europe	12.6	0.4	− 2.1	6.3
Japan	19.9	6.2	4.8	12.9

SOURCES: *Oil—the Present Situation and Future Prospects*, and *World Energy Outlook 1977*, OECD.

in the North Sea oil industry and the diminishing conventional reserves of oil and gas in North America. The rapidly increasing oil imports of North America, and of the United States in particular, contrast not only with the declining oil imports of Western Europe but also with policy declarations made by the United States government on the need to reduce dependence on imported oil.[10] A look at differences in governmental attitudes and policies will help explain the contrasting situations of North America and Western Europe with regard to oil imports.

First of all, there is the question of how the government bureaucracies of the United States and of most countries in Western Europe evaluate the long-term price elasticities of demand and supply for energy. As noted above, the neoclassical point of view seems to be more widespread in the United States, while the neo-Malthusian view is prevalent in Western Europe. This explains why many European governments have taken stronger action on energy. The ambitious energy-conservation and nuclear programs of France and West Germany are good examples.[11] Their relatively high dependence on imported oil may also be an important additional factor here.

Second, there is the question of how the political systems of

[10] U.S. Executive Office of the President, *The National Energy Plan*, Washington, D.C., 1977, p. 9ff.

[11] Horst Mendershausen, *Coping with the Oil Crisis*, The Johns Hopkins University Press, Baltimore, 1976, p. 67ff.

the United States and Western Europe function on energy issues. The implementation of ambitious energy policies in Western European countries is not only a result of government resolution but also a function of the relatively centralized political systems, which give governments a high degree of domestic control. Correspondingly, the problems of energy policy in the United States not only are a result of government indecision but also to a large extent derive from the decentralized political system, which leaves the government less control in domestic matters and gives economic and political interest groups considerable influence.[12]

The contrast also reflects different historical traditions. In Western Europe there is a long tradition of governmental control in economic life. This, in part, explains why there is less opposition within these countries to governmental initiatives in energy policy. However, in Western Europe governmental resolution in energy matters might eventually be undermined by opposition parties that are willing to accept greater dependence on imported oil in order to achieve faster economic growth and reduce unemployment. They might also receive the support of the opponents of nuclear programs as well as others who are dissatisfied with governmental energy and economic policies.

The outlook for the United States merits special attention because it is the world's largest oil importer and because changes in its oil imports are critical to the demand for oil on the world market. By comparing the energy situation of the United States with that of Western Europe and Japan we can still argue that the United States, at least theoretically, has considerable freedom of action in the relationship between economic policy and energy policy. Its combination of a high level of energy consumption in relation to population and gross national product (GNP) and large unused energy supplies makes it possible for the United States to pursue a relatively high rate of economic growth with declining imports of oil. In fact, about half of the OECD-area potential for cutting oil imports lies with energy

[12]Gabriel A. Almond, "A Comparative Study of Interest Groups and the Political Process," in Harry Eckstein and David E. Apter (eds.), *Comparative Politics*, The Free Press, New York, 1963, pp. 397–408.

conservation and expanded domestic energy production in the United States.[13] This means that American energy policy is of crucial importance to Western Europe and Japan. Put simply, the level of United States oil imports in many ways decides how much oil is left for the rest of the world.

The North American level of energy consumption in relation to GNP has traditionally been considerably higher than Western Europe's or Japan's.[14] Energy consumption in relation to GNP is now about 50 percent higher in the United States than in Western Europe, and almost twice as high as in France and West Germany. This reflects the historically great American abundance of energy that has fostered more energy-intensive patterns of production and consumption. The United States alone, with approximately 6 percent of the world's population, consumes about 33 percent of the world's energy.[15] One explanation is that since 1945 the rate of investment in the domestic economy has been much lower in the United States than in most West European countries; therefore there is a higher proportion of energy-intensive industrial equipment in the United States. However, most of the discrepancy can be accounted for by differences in patterns of consumption, primarily in the transportation sector and to a smaller extent in the residential and commercial sectors.[16] Essentially, American tastes for driving and temperature control make up most of the difference.

Over the past few years estimates of petroleum reserves and production potential in the United States have been revised downward. The *National Energy Outlook 1976* is not as optimistic about production potential as the 1974 *Project Independence Report*,[17] and the 1977 Congressional Research Service study, *Project Interdependence*, is even less optimistic.[18] The

[13]*World Energy Outlook*, p. 36.

[14]*The National Energy Plan*, p. 2.

[15]Ibid., p. 2.

[16]Darmstadter, Dunkerley, and Alterman, *Industrial Societies*, p. 25ff.

[17]U.S. Federal Energy Administration, *National Energy Outlook 1976*, Washington, D.C., 1976, p. 70ff.

[18]U.S. Library of Congress, *Project Interdependence: U.S. and World Energy Outlook Through 1990*, Washington, D.C., 1977, p. 15ff.

National Petroleum Council assumes in a study published late in 1976 that the potential for expanding reserves increases in direct proportion to the rise of crude oil prices.[19] However, production from existing fields is likely to decrease substantially over the next 10 to 15 years even with high oil prices. But at higher prices the decline of production from existing fields is likely to be less extreme than at current prices. Tertiary recovery, through thermal techniques and improved water and gas drives, could perhaps add 40 billion barrels, or 5.6 billion tons, to United States oil reserves.[20] Not all of these methods are technically feasible or commercially viable yet. Consequently, the oil production of the United States increasingly depends upon the discovery and development of new fields. A high rate of development will be needed to halt the decline in United States oil production. This new oil is likely to be quite expensive, implying high capital needs and much more capital mobilization for the energy sector.

Even with shrinking conventional oil supplies, it is difficult to argue that the United States is faced with a genuine shortage of energy. The country has large deposits of oil shale, coal, and uranium. In principle, these reserves could cover the energy consumption of the United States for a very long period of time.

Thus the United States has several alternatives for solving its energy problems. This freedom of action leads to difficult choices and political conflict over energy issues. The range of alternatives can be an obstacle to an effective energy policy, complicating the basic trade-off between conservation and an expansion of energy supplies.

From a purely economic point of view, there are ample opportunities for energy conservation in the United States because the level of energy consumption in relation to economic performance is considerably higher than in Western Europe or Japan.[21] In this comparative perspective conservation is the most sensible medium-term solution. Relatively small changes in patterns of gasoline consumption and in habits of temperature control would

[19]*Enhanced Oil Recovery*, National Petroleum Council, Washington, D.C., 1976, p. 27ff.
[20]*World Energy Outlook*, p. 41.
[21]Darmstadter, Dunkerley, and Alterman, *Industrial Societies*, p. 21ff.

conserve considerable amounts of energy and reduce oil imports. This point of view was also reflected in President Carter's national energy plan of 1977, which emphasized conservation more strongly than the expansion of domestic energy supplies.[22] In Carter's 1977 plan the key method proposed to stimulate conservation was to increase prices through taxation. The plan was well received in Western Europe[23] because, from a European point of view, it has been difficult not to associate the low energy prices in the United States with a certain amount of irresponsibility in an age of energy scarcity. In addition, rising American oil imports are often seen as a threat to Western Europe's economic stability.[24] Thus, both the aim, to reduce excessive energy consumption, and the method, taxation, appeared most reasonable to a West European public.

However, for historical and political reasons, energy realities are perceived differently in the United States. First, the United States has developed on a basis of inexpensive and abundant energy, which gave it a historical tradition of energy-intensive patterns of consumption. Mobility and temperature control are part of the "American dream." This is reflected in the decentralized pattern of settlement, low standards of insulation, and poorly developed public transportation systems. Second, because of the historically high level of energy consumption, there are important pressure groups that have an interest in maintaining past patterns of energy consumption rather than opting for energy conservation.

Thus, the President's energy plan inevitably has clashed with structural features and with strong vested interests. The ability of a price increase alone to check the growth of gasoline consumption seems dubious. In practice, demand for gasoline in the United States seems to be relatively unresponsive to gradual price changes.[25] In fact, demand seems to be affected more by

[22]*The National Energy Plan*, p. 25ff.

[23]*Le Monde*, April 22, 1977.

[24]Jean Carrié, "Les Incidences de la crise énergétique sur l'économie de l'Europe et des Etats-Unis," *Politique Etrangère*, no. 1, 1975, pp. 85–97.

[25]Verleger and Sheehan, "Demand for Gasoline," in Jorgensen, *Econometric Studies*, p. 213ff.

changes of income than by changes of price. It is also doubtful to what extent a price increase by itself can induce industry and, in particular, electric utilities to switch from oil to coal. The obtacles here are the environmental regulations on the use of coal, in particular on the sulphur content of emissions, and local resistance to coal mining.

In addition to price measures, a thorough reorganization of the transportation sector, including considerable investment, public subsidies, and perhaps direct regulations, might be needed. Possible measures include massive investments and subsidies for mass transit, a revitalization and reorganization of the railroad system, a mandatory program for air transport, and eventually a limitation on long-distance road transport of freight. Such measures are reasonable from a European point of view, but in the United States they imply profound changes and affect powerful vested interests. Thus it seems that a substantial reduction of gasoline consumption requires a transportation policy that is alien to American traditions and is politically divisive. In the absence of such comprehensive planning, though, it is possible that some of the energy-saving measures will lead in the long run to greater gasoline consumption, as savings to the public might lead to more driving by private individuals.

Given these prospects, energy issues are likely to be an important dimension of conflict in the American political system for a long time to come. The dispute over the 1977 energy plan essentially represents a confrontation between two different philosophies that has important consequences for the distribution of power. Two arguments against the 1977 energy plan are perhaps relevant. One is that Americans will maintain their habits of petroleum consumption at almost any price.[26] The other is that the taxation of energy consumption will bring about a higher degree of centralization of the United States economy, giving the federal government more power in economic matters and thus changing the overall balance of power in the American political system.

In this way, the battle on energy issues is not limited only to

[26]*Wall Street Journal*, November 1, 1977.

energy but extends to the issues of how the economy is organized and how political power is distributed. These broad implications may well reduce the ability of the American political system to work out a consistent energy policy, at least in the short and medium terms. Taxation implies a centralization of energy and economic decision making in the federal government and is based on the assumptions that the price elasticity of supply is low and that the prospects for expanding oil and gas output are limited, even at much higher prices. Giving the private sector control, by allowing the energy industry to raise prices and to benefit from tax breaks, presupposes that the price elasticity of energy supplies is sufficiently high to make market forces provide more energy, including more oil and gas. The weighing of these two points of view is likely to create political conflicts for years to come and to limit the ability of the United States to check the growth of oil imports.

Regardless of these political obstacles, the market seems to be making some contribution to energy conservation in the United States.[27] In 1976 and 1977, as economic growth picked up after the recession, the relationship between the rate of economic growth and the growth of energy consumption was considerably lower than in the past.[28] In fact, it seems that the energy coefficient of the United States has been reduced to approximately 0.6. This might indicate a profound change in patterns of energy consumption in the United States, mainly in response to the price increase, and in this case the freedom of action of the United States in economic and energy policy will have improved substantially.

However, it is difficult to draw conclusions on the basis of two years of experience. This decline in the energy coefficient may be cyclical rather than structural. Historically, the relationship between economic growth rates and the growth of energy consumption has varied considerably from one year to another. Also,

[27]*World Energy Outlook, April 1978*, Exxon Background Series, Exxon, New York, 1978, p. 8.
[28]"Energy: Where Did the Crisis Go," *New York Times*, April 14, 1978, Section 3, pp. 1, 9.

since 1973 the growth of the American economy seems to have been directed more toward the service sector than toward the industrial sector. Industrial production has expanded less than the GNP since 1973, which explains, to a certain extent, the low energy coefficients for 1976 and 1977. Finally, it is likely that the United States will have fairly low energy coefficients over the years when its conservation gains are achieved. But, after the level of energy consumption in relation to economic output has been reduced, energy coefficients could increase again when the potential for conservation has been realized.

Western Europe does not have the same freedom of action in dealing with the relationship between economic policy and energy policy. Historically, the economies of Western Europe have relied on scarce and relatively expensive energy. As a result, energy conservation has been taken seriously for a long time, and thus the European capacity to conserve energy while maintaining economic growth is relatively small compared with that of the United States. West European energy markets are already characterized by severe restrictions on consumption and the stimulation of production. The oil imports of Western Europe began to decline slightly in the mid-1970s because of North Sea oil and gas. European supplies of energy will expand further mainly through the development of nuclear power and North Sea petroleum. Some energy can be conserved in the industrial sector by investments in energy-saving equipment and changes in production processes. However, West European industry might argue that further energy conservation creates competitive disadvantages relative to American industry. A more important change in the relationship between energy consumption and economic growth could be achieved if growth in Western European economies shifts more substantially toward the service sector, but that implies a diminishing role for Western Europe as an industrial manufacturer.

Western Europe's energy record in the aftermath of the oil crisis is better than that of the United States.[29] Conservation efforts have been more substantial, and so has the expansion of

[29]*World Energy Outlook*, p. 45ff.

domestic supplies of energy. However, the ambitious nuclear programs have increasingly aroused political opposition, causing a slowdown in the expansion of nuclear energy. Also, the limited domestic supplies of hydrocarbons are somewhat insecure. In the Netherlands any expansion of gas production seems politically difficult, even if new reserves are found. In Great Britain a more restrictive depletion policy might be implemented when self-sufficiency in oil is attained (around 1980).

In Norway there might be considerable potential for increasing supplies of oil and gas.[30] Reserve estimates for the southern zone (south of 62°N) are conservative because relatively little drilling has taken place so far. The northern zone contains a large proportion of Western Europe's continental shelf, and geological data indicate the possibility of finding petroleum. If oil and gas are found there, the reserve estimates could be increased quite considerably and Norway might become a more important producer of oil in the late 1980s. The problem with the exploitation of these potential reserves is that Norway has limited economic needs and a small population that enjoys a high standard of living. Thus plans for substantially increasing production are likely to lead to serious political controversies. Events like the blowout in the North Sea in the spring of 1977 intensify such controversies. It is symptomatic that plans for exploratory drilling in the North were postponed afterward. However, Norwegian oil policy and particularly the low level of production could become major sources of friction between Norway and its allies and trading partners.[31] In addition, with a low rate of economic growth in Western Europe, Norway might produce more oil for economic reasons.

Of the major OECD countries, Japan has the least freedom of action in dealing with trade-offs between economic policy and energy policy. In 1975 oil imports represented 76 percent of Japan's total energy requirements, against 55 percent in Western Europe and 14 percent in North America. Even more than in

---

[30]Ibid., p. 46ff.

[31]U.S. Federal Energy Administration, *The Relationship of Oil Companies and Foreign Governments*, Washington, D.C., 1975, p. 143ff.

Western Europe, in Japan the economy has expanded on the basis of scarce and relatively expensive energy. This means that much of the conservation effort has already been carried out. As a result, the scope for conserving energy in the industrial sector through new equipment and changes in processes of production is limited. Also, industry might claim that strict measures of conservation lead to competitive disadvantages in international trade. The potential for expanding domestic energy supplies in Japan is almost exclusively linked to nuclear development.

OECD-AREA TENSIONS

The United States enjoys much greater flexibility than most of its allies and trading partners in the trade-off between economic policy and energy policy. Given the energy bind of the Europeans and the Japanese, a substantial increase in the oil imports of the United States is not only likely to cause economic problems for the entire OECD area, it will also produce substantial political divisions among them.[32] A close look at American policy on oil imports in the early 1970s demonstrates clearly the freedom of action of the United States and the anxiety this can generate in the OECD.

Before 1973 oil imports to the United States were controlled, and domestic oil prices were higher than on the world market. Consequently, energy prices were higher in the United States than in Western Europe or Japan, which was a competitive disadvantage for United States industry and an incentive for domestic oil production. This difference in prices could be eliminated if the United States lifted import controls and permitted domestic oil prices to move to the level of the world market, or if world market prices increased up to or beyond the level of American prices. The second possibility is not as implausible as it sounds; in fact, there are signs that from 1971 to 1973 the

[32]Seyom Brown, *New Forces in World Politics*, Brookings Institution, Washington, D.C., 1975, p. 36ff.

United States encouraged OPEC to raise its oil prices.[33] There are also signs that after the oil crisis the United States government discreetly supported the price radicals of OPEC, primarily Iran.[34] In addition to the desire to raise world prices, another reason for this stance could be that much of the financial surplus of OPEC was channeled to the United States. If the government of the United States really did encourage higher prices for oil after the oil crisis, this contrasts with two other important initiatives of the Nixon administration: Project Independence and the creation of the IEA. It might seem inconsistent that the Nixon administration would simultaneously launch an ambitious and expensive energy program, create an organization of oil consumers, and encourage OPEC to raise oil prices.

The contradiction here may be more apparent than real, and the link between these three facts could be logical. The Nixon administration could have proposed Project Independence to pacify domestic public opinion, knowing all along that the goal of energy self-sufficiency was unrealistic. The IEA could have been started to maintain American leadership, and to prevent the countries of Western Europe and Japan from making bilateral deals with OPEC countries. Finally, the Nixon administration could have encouraged high oil prices to maintain good relations with Iran, knowing that with higher world oil prices the competitiveness of American industry would improve.

This is why some Western Europeans have openly asked whether the United States under the Nixon and Ford administrations has played a two-sided game in oil politics, promoting its own interests at the expense of its allies.[35] The explanation is probably not so simple. The United States has different and sometimes very contradictory interests with regard to oil, and it therefore, perhaps, pursues policies related to oil that seem inconsistent. However, this explanation may not be sufficient to calm West European apprehensions.

[33]V. H. Oppenheim, "Why Oil Prices Go Up (1). The Past: We Pushed Them," *Foreign Policy*, no. 25 (1976–1977), pp. 24–57.

[34]"Pétrole: Les Tricheurs de Washington," *Le Nouvel Observateur*, no. 632, 1976, pp. 36–37.

[35]Ibid., p. 36.

Oil imports function as a safety valve in the United States energy sector; importing oil is the least controversial short-run solution to American problems of energy policy. But greater United States oil imports will increase the demand for oil on the world market, and this in turn will encourage OPEC to make new increases in the price of oil. Given these circumstances, apprehensive Western Europeans ask whether there is a plot behind the increasing oil imports of the United States.[36] They argue that the United States has a real interest in increasing its oil imports in a medium-term perspective. By maintaining price controls on oil and gas, and letting oil imports increase, the United States achieves the following:

- An average level of energy prices below that of Western Europe or Japan, which gives a competitive advantage to American industry
- A slower rate of depletion of domestic oil and gas reserves, which means greater future reserves
- Eventual pressures on the world oil market, which lead to price increases that make the exploitation of alternative sources of energy economical and put the United States in a more favorable position, given its technology and resources

It is perhaps doubtful that the decentralized political and financial system of the United States would permit such a calculated manipulation of the situation. But this does not rule out the possibility that the present stalemate in United States energy policy, and the rising oil imports resulting from it, could work out to be in the long-term interest of the United States. Therefore, West European fears are quite legitimate. In a tighter supply siutation this could be a major source of friction between the United States and most of its allies. One symptom is that already considerable pressure is being put on the United States in international forums, such as OECD and IEA. The Secretary General of the IEA has recently stated that the ability of the Western world to

[36]"L'Absence de politique énergétique américaine fait le jeu de l'OPEP," *Le Monde*, March 23, 1976.

TABLE 13
The Consumption of Various Types of Primary Energy in 1976, in Millions of Metric Tons of Oil or the Equivalent

	Total	Coal	Hydro-electric	Nuclear	Natural gas	Oil	Oil imports
North America	1,998	365	113	46	580	894	354
Western Europe	1,208	235	74	27	175	697	607
Japan	346	52	21	7	11	255	254

SOURCE: *Energy Statistics 1974–76*, OECD, Paris, 1977.

limit its reliance upon imported oil depends on the United States.[37] This can be interpreted in two ways. First, the size of the United States makes its oil imports decisive, and second, its example is crucial. If the United States does not promote energy conservation and an expansion of domestic energy supplies, most countries of Western Europe and Japan will have stronger reasons to go ahead with conservation and nuclear programs, but chances are that the psychological impact of the United States will slow down energy programs in other OECD countries. Ultimately these differences can only lead to increasing friction between the governments.

It can easily be argued by Japan or Western Europe that for them oil imports are a vital necessity for economic life but for the United States they are a luxury, an excess consumption of energy. An oil cutback for Japan and Western Europe would lead to a total disruption of economic life, whereas for the United States it probably would only mean restrictions on private driving.[38]

A closer look at the energy balances should make this clear (see Table 13).

[37]Ulf Lantzke, "IEA Head: U.S. Key to Energy Economy," *Oil and Gas Journal*, April 18, 1977, pp. 25–26.
[38]Jean Carrié, *Les Incidences de la crise énergétique*, pp. 85–97.

In North America, oil imports represent only 18 percent of total energy consumption. For Western Europe the proportion is 51 percent, and for Japan 74 percent.

TABLE 14
End Uses for Oil for Selected Sectors in 1974, in Millions of Metric Tons

	Total for all sectors	Industry	Transportation	Road transport
North America	777	98	428	358
Western Europe	667	141	158	121
Japan	238	61	38	29

SOURCE: *Energy Statistics 1973–75*, OECD, Paris, 1977, pp. 76–77, 88–89, 100–101, 106–107, 112–113.

In North America the total oil consumption of the transportation sector alone exceeds the volume of oil imports (see Table 14). In Western Europe and Japan the oil consumption in transportation represents only a small proportion of the oil imports. In 1974 the total oil consumption in the OECD area was 1,781 million tons, of which 639 million tons, or 36 percent, were consumed by the transportation sectors. Road transport alone took 530 million tons, or 30 percent. It should be noted that the transportation sector of the United States took 394 million tons, and United States road transport alone 330 million tons. This corresponded to 19 percent of the total oil consumption of the OECD area, and to 28 percent of the OECD oil imports.

OECD-AREA OPTIONS

The range of possibilities for the OECD countries as a group is illustrated on the theoretical level by the future relationships between three basic economic components that determine the demand for oil imports. These are the rate of economic growth, the energy coefficient (the relationship between the growth of energy consumption and the rate of economic growth), and the growth of domestic energy production. Table 15 gives the net oil imports to the OECD area by 1985 according to different combinations of these three basic components.

On paper the OECD area has considerable freedom of action in reconciling economic and energy policies. It can maintain a

TABLE 15
OECD Oil Imports by 1985

Energy coefficient	Growth of domestic energy production (%)	Rate of economic growth			
		2 %	3 %	4 %	5 %
1.1	2.5	1,426	1,977	2,590	3,270
	3.0	1,262	1,813	2,462	3,106
	3.5	996	1,641	2,254	2,934
	4.0	909	1,460	2,073	2,753
1.0	2.5	1,168	1,821	2,359	2,952
	3.0	1,004	1,657	2,195	2,788
	3.5	832	1,485	2,032	2,616
	4.0	651	1,304	1,842	2,435
0.9	2.5	1,240	1,669	2,138	2,649
	3.0	1,076	1,505	1,972	2,485
	3.5	904	1,333	1,802	2,313
	4.0	732	1,152	1,621	2,132
0.8	2.5	1,150	1,522	1,924	2,359
	3.0	986	1,358	1,760	2,195
	3.5	814	1,186	1,588	2,023
	4.0	633	1,005	1,407	1,842
0.7	2.5	1,043	1,379	1,719	2,086
	3.0	879	1,215	1,555	1,922
	3.5	707	1,043	1,383	1,750
	4.0	526	862	1,202	1,569

SOURCE: These figures were derived by extrapolating from the 1975 figures.

high rate of economic growth without overly increasing oil imports if appropriate policy measures are taken. For example, a rate of economic growth of 5 percent could by 1985 give only a modest growth in oil imports, provided the energy coefficient falls to 0.7 and the domestic energy production expands at 4 percent a year. Yet a relatively low rate of economic growth could create a considerable increase in oil imports, if energy policy is unchanged. For example, a rate of economic growth of 3.0 percent and an energy coefficient of 1.0 combined with an expansion of domestic supplies of energy at a rate of 2.5 percent a year would require oil imports to be 50 percent higher in 1985 than in 1975.

Even if the examples chosen here are extreme, the comparison shows that there is a certain flexibility in the OECD area as a whole in dealing with the trade-offs between economic policy and energy policy. The major problem is the economic and political costs to governments of energy policy measures. In reality, most OECD governments have the choice between the following options in the relationship between economic policy and energy policy:

- Sustaining economic growth, with a limited effort in the energy sector, thus increasing oil imports.

- Sustaining economic growth, with a significant effort in the energy sector to check the growth of oil imports. Such an effort is likely to be economically costly and to provoke serious political opposition. Its potential for success is higher in the United States than in most other OECD countries, given the greater potential for conservation and for increasing domestic energy supplies.

- Restricting economic growth in order to check oil imports.

The first option preserves economic growth and thus satisfies most domestic goals, but at the cost of increasing oil imports and international vulnerability. The second and third options decrease national exposure to the world oil market but impose domestic costs in the form of an all-out energy policy or economic stagnation.

As noted above, the energy policy of the United States, through its impact upon the level of oil imports, is a decisive factor in the energy choices and economic policies of most other OECD countries.[39] Consequently, increasing United States oil imports will not only put pressure on the American balance of trade, making the United States more vulnerable to diplomatic pressure from oil-exporting countries, but will also mean more political pressure from Western Europe and Japan. These pressures, however, function to a considerable extent as "externalities" in the American political system. They are perceived as problems by those dealing with foreign relations, but most of the United States economy is sheltered from them, and thus their political weight is limited. Therefore, growing external pressure could certainly lead to stronger initiatives on energy policy by the executive, but because of domestic opposition the chances of these programs becoming law will be fairly limited unless a major international energy crisis develops.

Clearly, the future prospects for oil supplies on the international market are a decisive factor for the formulation of energy policy in the United States and in most other OECD countries. Therefore, a look at the OPEC response to OECD needs is crucial to a complete picture of the world oil market in the medium-term future.

[39]Guy de Carmoy, *Energy for Europe*, American Enterprise Institute for Public Policy Research, Washington, D.C., 1977, p. 96ff.

FOUR

The OPEC Response

THE HISTORICAL BACKGROUND

From 1960 to 1973 oil production in the OPEC countries increased from 433 to 1,307 million metric tons, an average yearly growth rate of 8.9 percent. Even with a low oil price and government income a relatively modest fraction of total revenues, the expanding oil industry had profound economic and social effects in the OPEC countries. It is clear that, without the presence of the oil industry, the OPEC countries would not have reached their present levels of development.[1] The oil industry accelerated a process of economic and social change characterized by increasing industrialization, urbanization, and education. By stimulating this process, the First Oil Regime prepared the way for a transition to a new regime. In the OPEC countries foreign control of the oil industry and modest income in relation to the total oil rent—the sum of all profits and other incomes from oil—were increasingly felt to be unjust. In many cases, the rapid extraction of finite national resources was seen as economically irresponsible given the future needs of the domestic economy. Foreign control of the oil industry was essentially felt to be morally and politically degrading.

[1]Edith Penrose, "Aspects of Consumer/Producer Relationships in the Oil Industry," in Ragaei El Mallakh and Carl McGuire (eds.), *U.S. and World Energy Resources: Prospects and Priorities*, ICEED, Boulder, Colo., 1977, p. 23ff.

It is important to underline at this point the recent colonial or semicolonial past of practically all OPEC countries. They either had been under direct foreign control, as in the cases of Iraq and Algeria, or had been dominated economically by foreigners, which limited their national sovereignty, as in Iran and Venezuela. An important exception is perhaps the United States–Saudi relationship, which avoided many problems experienced by others.[2] Foreign control of the oil industry was thus seen as a colonial or semicolonial legacy. As recently as during the Six-Day War in 1967 the Arab oil exporters were unable to use oil for political leverage. The OPEC countries felt that they were being exploited both economically and politically and that foreign control of their oil industries was incompatible with important goals of economic policy, foreign policy, and national dignity. As in other Third World countries, control of the basic national industry was seen as an important step on the road to independence.[3]

Despite these similarities, OPEC is in many ways a unique phenomenon. Historically, most cartels of raw-material producers have been unstable and have collapsed after a few years because of a failure to regulate production and distribute income equitably.[4] Excess capacity has led to underbidding, the disruption of the cartel, and finally price regulation through the market.[5] OPEC is special for several reasons. First, the price elasticity of demand for oil is very low, and oil has a particularly limited resource base. Furthermore, the excess capacity for production is very unevenly distributed, a point that will be examined in detail below. The oil-producers' cartel also functions in a relatively flexible way. In fact, OPEC only sets the price of oil, leaving questions of production and export levels to member countries. This procedure allows some flexibility and frees OPEC from many of the tensions that have plagued other exporter

[2]David E. Long, *Saudi Arabia*, Center for Strategic and International Studies, Georgetown University, Washington, D.C., 1976, p. 18ff.

[3]Penrose, "Consumer/Producer Relationships," p. 22.

[4]Paul Leo Eckbo, *The Future of World Oil*, Ballinger, Cambridge, Mass., 1976, p. 2ff.

[5]Ibid., p. 25ff.

cartels. The price question can, however, be divisive, as was demonstrated at the OPEC meeting in Doha in December 1976 when Saudi Arabia wanted a smaller price increase than the other members of the cartel.

Politically, OPEC can be described as an international interest group or a trade union of raw-material producers. In addition to its function as a price fixer, OPEC is also a forum for political discussions and a platform for common demands.[6] All OPEC countries share the success of the organization. Their improved standing in the world is closely related to the performance of OPEC, and this makes it important for them to maintain political cohesion and solidarity in spite of obvious differences in interests and points of view. There is as a result a stronger ideological cohesion in OPEC than in most, if not all, other cartels of raw materials exporters in recent history. This contributes to OPEC's chances for survival and explains why minor price disputes pose no immediate threat to the cartel.

The crucial question is: To what extent does the oil revolution mark a historic break in oil supply trends? The OPEC countries now get a much better price for their oil and have control of production. From 1973 to 1975 OPEC oil exports declined, absorbing the decline in OECD demand. The reduction of output and exports was due more to the evaluation by each government of market conditions than to any concerted action. After 1975 OPEC oil exports increased again, in response to OECD demand.

ECONOMIC AND POLITICAL EXPLANATIONS

The oil exports of the OPEC countries can be understood from economic as well as political perspectives. In economic terms, the oil exports of the OPEC countries are the response of an oligopolistic cartel of producers to demand on the market. Politically, the oil exports of the OPEC countries are the result of policies and budgetary decisions formulated in the context of national and international politics by a limited number of gov-

[6]Grethe Vaernoe, "OPEC—Kartell eller fagforening?" *Samtiden*, no. 2, 1975, pp. 65–78.

ernments. These two points of view are complementary, but for practical reasons they will be presented separately.

The main economic factors behind oil exports are the income requirements of the various OPEC countries, the size of their oil reserves, and the market conditions. The oligopolistic position enables the cartel, or its dominant members, to control the price of oil through decisions on the volume of exports. An oligopoly has two economic goals: to maximize its income in the short run and to maintain its position over time. Therefore, it is rational economic behavior for oligopolists to create a price that gives them an optimal income. This can be defined as the greatest possible income compatible with continued demand on the market and with the maintenance of a controlling position by suppliers. Too high an income might hurt demand, the position of the oligopoly, and thus future income. Too low an income could imply an underutilization of income-earning opportunities.

In determining the volume of exports, the individual cartel members have to take into account not only their own income requirements and long-term interests but also market conditions, especially the relationship between the amount they supply and the price this creates. The responsiveness of demand to changes in price is the crucial factor here. So OPEC's decision-making process might be described dynamically as follows. The OPEC members know they have certain income needs, and they attempt to maximize their incomes to meet these needs. Given their oligopolistic position as price setters, they must strike a delicate balance, creating a price that maximizes their income without straining the consumers in such a way that future demand declines or their oligopolistic position is undermined by alternative energy sources.

The main political factors weighed in considering the price and volume of oil exports from OPEC countries are the goals of economic policies, the long-term political interests related to oil, and foreign policy concerns. Its oligopolistic position enables OPEC to use oil exports for leverage in foreign policy. Its members seek both to maximize influence in the short term and to maintain their position over time. This calls for a certain amount of compromise because exerting strong influence in the short run could provoke countermeasures that limit their future leverage.

Thus, in deciding the volume and directions of exports, individual OPEC countries have to take into account not only their own interests and policy goals but also international political conditions and, in particular, the sensitivity and vulnerability of oil importers to political pressure.

The economic approach stresses the importance of the market, and the political approach emphasizes policy decisions and national interests. Both underline the dominant position of OPEC in the world oil market. Relatively small changes in the economic requirements of individual members or in their political perceptions can have quite important effects on the volume, price, and possibly the pattern of oil trade.

National control of their oil industries and their rate of output has enabled the OPEC countries to achieve a better balance between oil conservation and their income requirements. Some countries, in particular Kuwait and Libya, have reduced output of oil for reasons of conservation. The financial wealth created by the oil revolution has accelerated the process of economic and social change initiated under the First Oil Regime and has stimulated ambitious programs of economic development. But for many years oil is going to be the basic national asset of the OPEC countries. This implies a delicate trade-off between income requirements, political interests, and conservation policies.

Among the OPEC countries there is a wide range of national diversity despite the basic generalizations that have been made about their economic and political positions as a group. These national differences determine the way a given country behaves. The best way to comprehend the spectrum of OPEC behavior is to examine the motivations of the various countries that constitute the cartel.

CONTRASTING POSITIONS OF OPEC MEMBERS

A major feature of OPEC is the uneven distribution of oil reserves and population among its members. OPEC can be divided into two groups: One group has small reserves in relation to population, and the other has large reserves in relation to population (see Table 16).

TABLE 16
Oil Reserves and Population of OPEC Countries 1977

Country	Reserves 1977 (millions of metric tons)	Population 1977 (thousand)	Ratio of reserves to population
Algeria	932	17,790	0.05
Ecuador	233	6,705	0.03
Gabon	291	535	0.54
Indonesia	1,438	134,414	0.01
Iran	8,631	34,424	0.25
Iraq	4,658	11,759	0.40
Libya	3,494	2,607	1.34
Nigeria	2,672	80,000	0.03
Venezuela	2,092	12,542	0.17
Sum group I	24,441	300,776	0.08
Kuwait	9,234	1,112	8.30
Qatar	781	183	4.26
Saudi Arabia	20,550	9,430	2.17
United Arab Emirates	5,017	652	7.69
Sum group II	35,582	11,377	3.13
Sum OPEC	60,023	312,153	0.19

SOURCE: *International Petroleum Encyclopedia, 1977*, p. 280, and Arthur S. Banks (ed.), *Political Handbook of the World: 1977*, McGraw-Hill, New York, 1977, pp. 7–483 passim.

The countries of the first group, because of their small reserves, have relatively short production profiles, with proven oil reserves corresponding to about 30 times the 1975 level of output (see Table 17). They cannot increase their output in order, for example, to finance more imports without further shortening their production profiles. Yet, because of their development programs, their income requirements will, by the early 1980s, probably make it difficult for them to tolerate a loss of income. However,

TABLE 17
OPEC Oil Production and Oil Reserves in 1975,
in Millions of Metric Tons

Country	Production	Reserves	Ratio of reserves to production	Possible ceiling 1985*
Algeria	48	1,053	21.9	50
Ecuador	8	342	42.7	10
Gabon	11	239	21.7	25
Indonesia	65	2,051	31.5	100
Iran	267	9,028	33.8	320
Iraq	111	4,787	43.1	150
Libya	75	3,638	48.5	120
Nigeria	89	2,859	32.1	120
Venezuela	117	2,051	17.5	100
Sum group I	791	24,995	31.5	995
Kuwait	104	11,141	107.1	120
Qatar	22	820	37.2	35
Saudi Arabia	353	23,686	67.0	1,000
United Arab Emirates	84	4,640	55.2	100
Sum group II	563	40,287	71.5	1,255
Sum OPEC	1,354	65,286	48.2	2,250

*Author's estimates.
SOURCE: *Basic Petroleum Data Book*, section XIV, tables 1 and 2.

their income may drop if they must reduce exports in order, for example, to keep the oil price high. The countries with large oil reserves, the second group, have relatively long production profiles. Their proven reserves on the average correspond to 71 times their 1975 levels of output. Thus they can increase production in order, for example, to achieve a specific relationship between supply and demand, without having to confront a short production profile. They can also afford to reduce output, for

TABLE 18
OPEC Oil Production and Foreign Transactions in 1975 with Estimates of Future Production Limits, in Millions of Metric Tons of Oil and 1975 U.S. Dollars

Country	Oil production	Oil consumption	Oil exports	Government revenue per ton ($)	In billions of dollars				Oil production limits		
					Other exports	Imports	Grants	Assets	1980	1985	1990
GROUP I											
Algeria	47.50	0.50	45.00	91	1.21	7.08		1.50	50	55	40
Ecuador	8.00	2.50	5.50		0.56	1.29		0.30	10	10	10
Gabon	10.00	0.25	9.75	45	0.27	0.75		0.12	12	12	12
Indonesia	66.00	9.65	56.25	69	1.90	7.05		0.83	100	90	80
Iran	265.00	16.00	249.00	77	2.44	19.44		14.95*	320	290	250
Iraq	112.50	6.50	106.00	79	0.43	6.90		4.34	160	180	200
Libya	74.50	2.00	72.50	82	0.31	6.04		2.67	100	100	90
Nigeria	89.25	3.00	86.25	81	0.98	8.07		6.00	105	105	100
Venezuela	120.00	12.00	108.00	77	1.02	7.07		8.00	110	100	80

GROUP II

Kuwait	105.00	1.35	103.65	77	0.82	2.90	1.00	14.43	110	100	80
Qatar	22.00	0.25	21.75	81	0.04	0.88	0.09	2.83	24	24	24
Saudi Arabia	354.00	8.50	346.50	75	1.52	11.29	1.00	43.20	750†	1,000†	1,250†
United Arab Emirates	85.00	0.40	84.60	79	0.75	3.99	1.00	7.16	105	100	100

*Taking into account debt payments of $1.0 billion a year.

†These indicate the range of alternative production limits for Saudi Arabia.

SOURCE: Alexander Caldwell of the Crocker National Bank provided these data.

purposes of conservation or to manipulate prices, because of their limited income requirements and their extensive financial assets.

Thus the two groups confront different situations. One group has large income requirements and little ability to regulate its level of output. The other has limited income requirements and considerable freedom of action in regulating oil-production levels. Consequently, the first group has little flexibility in dealing with choices between economic policy and oil policy, while the second group has a variety of options.

This contrast can be modified by a closer examination of the situation in particular countries. In the first group Libya and Iraq have the largest reserves in relation to population (after Gabon), the longest production profiles, and the largest surpluses on the 1975 balance of trade. In the second group, Saudi Arabia has an exceptional position, with the largest oil reserves in the world, the longest production profile (after Kuwait), the largest surplus on the balance of trade, and important financial assets. Saudi Arabia can thus legitimately be treated as a special case. The combination of large physical and financial resources with a small population gives Saudi Arabia an extraordinary ability to regulate its output and export levels. Saudi Arabia has the ability to flood the market with oil or cut off exports entirely for a prolonged period of time. Consequently, investment in additional production capacity can be seen as increased political bargaining power, rather than as simply the intent to produce more. The accumulation of financial assets can also be seen as a source of political bargaining power. This combination of flexibility and market predominance gives Saudi Arabia tremendous economic and political clout, particularly in the short run.

The unequal distribution of oil reserves and of population also creates different economic interests for the two groups of countries. Because of the short production profiles of its members, the first group has an interest in maximizing income in the short run rather than perpetuating the long-run position of the cartel. The second group has an interest in maintaining the long-run position of oil because of the long production profiles of its members. Thus the two groups have strikingly different views of the optimal income from oil. The countries of the first group are developing alternative sources of income because their oil reserves will not last long. They therefore have a natural interest in obtaining a high price for their oil now in order to finance an accelerated program of industrialization. The countries of the second group are not likely to develop alternative sources of income at home on a large scale very soon, and they will probably depend on their oil exports for several generations to come. However, increasing financial investment abroad, mainly in the OECD countries, does provide an additional and in many ways alternative source of income. Thus the countries of the second group have a strategic interest in keeping the price of oil at a level where it can find buyers and remain competitive with alternative sources of energy.

To sum up, the OPEC countries seem to fall into two groups and one special case on the basis of their physical endowments and economic interests.

- *Group I*
  Countries with small reserves and relatively large populations, production profiles largely less than 30 years, ambitious plans for economic development, large income requirements, an interest in maximizing oil income in the short run, and an interest in obtaining maximum short-run political influence from oil

- *Group II*
  Countries with larger reserves and relatively small populations, production profiles of more than 50 years, fewer possibilities to develop alternative sources of income, moderate present income requirements, an interest in maintaining a market for oil over the long run, a desire to maintain their long-run political influence derived from oil, and an increasing interest in preserving the position of their investments abroad, especially in the OECD countries

- *Saudi Arabia*
  Having the world's largest oil reserves and being the world's largest exporter of oil, this country has an interest in main-

taining a market for oil over the long run, maintaining its long-run political influence derived from oil, retaining its position as the world's largest oil exporter, conserving its oil reserves, and keeping the world economy in a state that protects its own financial assets

The different resource endowments and economic interests within OPEC quite naturally provide the basis for serious political differences over oil policy. To a certain extent these political differences relate directly to the economic and geographic factors described above. Countries with a relative abundance of oil are often more conservative and want to preserve their position. Their counterparts, with relatively little oil, tend to be radical and desire to use their influence while they can. There is, however, more to the divergent views OPEC members have of the optimal political influence they can obtain from oil, and these differences do not quite parallel their economic differences. They are, in fact, more complex.

The broad spectrum of interests in OPEC politics creates an array of factions within the cartel. For example, among the members of OPEC with a relative abundance of oil reserves, there are both radicals and conservatives on the price question. Saudi Arabia, which in a sense stands apart because of its predominance, can also be seen as the key conservative state. Along with Qatar and the United Arab Emirates (UAE) it has so far favored stable oil prices and is cautious about using oil for short-term political leverage. Kuwait, which in economic terms is also a member of the same group, often sides on price disputes with radicals like Algeria and Iraq. These states, in contrast to the conservatives, seek higher prices and are more aggressive in the use of the political influence derived from oil. Unlike the countries with relatively small reserves, Kuwait has more influence on prices and export levels because of its extra flexibility in controlling supplies.

The unequal distribution of oil reserves between the two groups of countries indicates a gradual shift of production away from non-Arab countries. This shift will make Arab politics even more relevant to oil supplies and oil prices, and it will strengthen the position of the Organization of Arab Petroleum Exporting Countries (OAPEC). OAPEC was created much later than OPEC, in 1968, in order to provide closer cooperation between the Arab oil exporters. Even if OAPEC does not formally interfere with the functioning of OPEC, it is clear that the gradual shift of production will have important consequences for relations between the two organizations.

Israel and the Middle East conflict in general are of primary importance to the Arab states, and this creates further divisions within the cartel. OAPEC is not known for its internal cohesion, but it does represent a formidable block of oil exporters that take a hard line toward Israel. The Arabs, with the exception of Iraq, worked together in imposing the 1973–1974 oil embargo, while non-Arab states continued to export oil at the same level (Venezuela) or at slightly higher levels (Iran, Nigeria, and Indonesia). Iran maintained oil exports to Israel during the embargo but at the same time pressed very hard for a price increase. The softer positions taken by non-Arabs during the embargo are a reflection of their desire for revenues as well as their stance toward Israel. The difference between Arabs and non-Arabs over Israel emphasizes the cultural splits within OPEC. Nigeria, Venezuela, Iran, Indonesia, and the Arab countries are culturally quite divergent and can be expected to behave differently in international politics as a result. Strategically there are also definite cleavages. Kuwait and the small Gulf states are militarily vulnerable, while Iran has built itself up to be the dominant power of the region and a critical element in OPEC because of its control of the Gulf. The common interest in oil and the success of OPEC help prevent any of these differences from breaking up the cartel, but disagreements do obviously create strains in its operation.

Looking to the future, potential discord among OPEC countries, over issues not directly linked to oil, will involve the conflict in the Middle East and the development of the Third World. The Arab OPEC countries have a direct stake in the outcome of the Arab-Israeli conflict and might desire a high degree of flexibility in questions of oil prices and oil supplies to be able to create an appropriate mix of moderation and toughness for political purposes. But this policy could strain relations with the non-

Arab OPEC countries, which might desire more structured control over oil prices. All the non-Arab member countries have large populations and small oil reserves, and they can profit from the political use of oil only when it takes the form of a price hike, as was the case in 1973–1974.

Saudi Arabia confronts a dilemma if the Arab-Israeli conflict is settled. At present, by keeping the price of oil down and guaranteeing supplies, it can exercise leverage over the United States. This can be called the negative use of the oil weapon, and the policy was explicitly expressed at the OPEC meeting in December 1976.[7] If the Arab-Israeli conflict is settled, perhaps as a result of American influence over Israel, Saudi Arabia would lose the force of its argument to keep prices down or even to guarantee supplies. In such a situation, Saudi Arabia would be more open to pressures from other OPEC countries, as well as from domestic groups wanting a change of oil policy. Thus a settlement of the Arab-Israeli conflict might be beneficial to the other OPEC countries, whereas a continued stalemate with tensions being gradually reduced might, at least objectively, be more beneficial to the present Saudi government and perhaps to the consuming countries as well.

Future splits over policy toward the Third World could develop between OPEC countries wanting to keep the price of oil down for the benefit of the Third World and those seeking to assist the Third World with funds accumulated through a higher oil price. Saudi Arabia could use the balance of payments of most Third World countries as a pretext for refusing to increase the price of oil and possibly try to gain political support from countries of the Third World in order to offset its possible isolation in the Middle East.[8] Other OPEC countries, desiring higher prices, might propose to finance generous programs of economic aid from the greater revenues coming from the oil price increase, perhaps offsetting the increased burden on the Third World. The latter solution would give OPEC countries more direct political influence in the Third World.

[7]Louis Turner, "Oil and the North-South Dialogue," *The World Today*, February 1977, pp. 59–60.
[8]Ibid., p. 59.

The newfound power of OPEC not only has created divisions in the cartel but also has had ramifications for the structure of international relations. In the preceding chapter we saw the impact of oil politics on the OECD area; now we turn briefly to North-South relations, the less developed countries, and the Eastern bloc.

OTHER ACTORS

In the context of North-South relations there is an amazing unity between oil producers and the less developed countries that was exemplified at the Conference on International Economic Cooperation held in Paris in 1976–1977. The alliance seems to contradict the economic interests of the LDCs because, like the OECD countries, they have also suffered from increased oil prices. The unity results from the shared colonial heritage of the two groups and the need for political compromise. The LDCs have put up with the higher oil prices and have taken a hard stand against Israel in international bodies such as the United Nations General Assembly and the United Nations Educational, Scientific, and Cultural Organization. This solidarity with OPEC interests has been recognized by the cartel in its refusal to separate negotiations on oil from issues of development and commodity agreements in talks with the North. To both OPEC countries and LDCs, the UNCTAD (United Nations Conference on Trade and Development) bloc provides a wider forum for their demands.[9]

In terms of the actual volume of oil trade, less developed countries have so far been of little significance, but they clearly have an important political role. Their future unity with the oil producers on Israel and the North-South question depends in part on their economic prospects. The increase in the price of oil has strained the balance-of-payments positions of many developing countries and has also stimulated oil exploration and production. A good deal of the world's search for oil goes on in

[9]Penrose, "Consumer/Producer Relationships," p. 24.

less developed countries. A few are even becoming net oil exporters, including Egypt and Mexico and possibly Malaysia and Vietnam. Countries like Brazil and India cover an increasing part of their own needs. Net oil exporters could in many cases join OPEC.[10] Still, the vast majority of developing countries are dependent on imported oil. With increasing industrialization their oil imports are likely to grow quite quickly.

The ability of these importers to pay for foreign oil differs dramatically. Some Third World countries are advanced enough or have an adequate export earning potential to pay for most of the oil they need. Others, mostly in the Fourth World, are unlikely to be able to pay for their energy imports for a long time to come. In these cases the risks of economic setbacks and intolerable debt burdens are real and pose a threat to the political unity of OPEC and the South. As the oil and energy problems of these less developed countries become more acute, OPEC will increasingly have to consider their demands.

According to their energy needs the various members of the Third World will probably respond to higher OPEC prices by some combination of the following policies:

- Trying to use the alliance with the OPEC countries to get economic advantages from the OECD countries[11]

- Trying to get special agreements with OPEC countries for a reduced oil price, credit, or financial compensation through aid or OPEC investment

- Siding with OECD countries against OPEC in the hope of preventing future price increases or getting aid from them to compensate for oil costs

- Attempting, in some instances, to clear their debt burdens through deliberate default

- Forming their own cartels and interest groups to put more pressure behind their demands

[10]For example, several minor oil producers, such as Trinidad, have already applied for membership to OPEC.
[11]Penrose, "Consumer/Producer Relationships," p. 23ff.

• Opting for crash programs of energy production if they can afford it

With regard to this last point, nuclear power does provide one solution for relatively wealthy LDCs, but it is also a source of potential instability. A few LDCs may acquire nuclear technology not only as a solution to their energy problems but also as a means of obtaining nuclear weapons. They could thus increase their political prestige and influence in the world, at the risk of greater international political instability.

Both China and the Soviet Union have followed OPEC pricing. China has considerable potential as a producer of oil. Its production of oil has been increasing quickly over the past years, reaching in 1976 perhaps 91 million tons. Even if production should continue to grow at high rates, the growth of domestic consumption is likely to limit the quantity of oil available for export.[12] Thus, in the 1980s, China is not likely to be an important competitor with OPEC in oil exports to Far Eastern markets. Use of foreign capital and technology on a massive scale, via joint ventures and other means, could significantly increase prospects for Chinese oil exports, but this solution seems politically unacceptable.[13]

The Soviet Union is now the world's largest producer of oil. Its oil production has had an exponential growth quite similar to that of the United States but with a certain time lag. Soviet oil production now seems to be running into some of the same problems already faced in the United States. In the traditional areas of production the oil reserves are being depleted, and production has been moving into steadily less accessible areas, mainly in Siberia. Here, technical problems are considerable, creating high costs and long lead times. Other bottlenecks are shortages of skilled labor and inadequate transportation facilities.[14] Therefore, past growth rates for Soviet oil production are

[12]U.S. Library of Congress, *Project Interdependence*, Washington, D.C., 1977, p. 73.
[13]Ibid., p. 73.
[14]Ibid., p. 67ff.

unlikely to continue, and it is even questionable whether present plan targets are realizable. Nonetheless, Soviet domestic oil consumption is likely to continue to grow at relatively high rates.[15] This means that sometime in the 1980s the Soviet Union is likely to face the choice of curtailing domestic consumption, reducing exports to Eastern Europe and the OECD area, or counting on Western capital and technology in order to expand oil production.

Each of these options has undesirable side effects for the Soviet Union. Curtailing domestic consumption could slow down agricultural improvements, petrochemical expansion, and the growth of the consumer goods sector. Reducing oil exports to Eastern Europe implies the partial loss of an instrument of political control. Eastern Europe has traditionally been dependent on Soviet oil, but in recent years it has been clear that Soviet oil supplies would not be expanded. As a result, Eastern Europe is increasingly coming into the world oil market as an importer. Greater oil imports from outside the Comecon area must be paid for by a growth in exports, and this can be an important factor in the relative decline of inter-Comecon trade, which could adversely affect political cohesion in the area.

Reducing oil exports to OECD countries will limit Soviet foreign exchange earnings, reducing their ability to pay for imports of Western technology, grain, and other items.[16] The use of Western capital and technology, for example, in the form of joint ventures, could improve Soviet oil production but might also require close cooperation with the West, which would tie the Soviet Union to the OECD area.[17] This might not be politically acceptable to the Soviet Union.

Some sources argue that the exponential growth of Soviet oil production is likely to peak in the 1980s and that the Soviet Union is then likely to become a net importer of oil.[18] If such a situation occurs, it might bring about a substantial price increase for OPEC

[15]Ibid.

[16]U.S. Congress, Senate, Committee on Interior and Insular Affairs, *Geopolitics of Energy*, 95th Cong., 1st Sess., 1977, p. 16ff.

[17]*Project Interdependence*, p. 70.

[18]Jahn Otto Johansen, "Sovjetisk olje-og gasspolitikk," *Internasjonal politikk*, no. 2, 1976, pp. 201–223.

oil.[19] It might also lead to competition between the Soviet Union and the United States for Middle East oil. This certainly would not contribute to stability in the area or on the world oil market. Given these prospects, it is reasonable to assume that the Soviet leadership will give priority to finding and developing new Soviet oil fields, even at very high costs, to avoid becoming a net importer.

POLICY OPTIONS IN OPEC

The unequal distribution of resources and populations is an important conflict dimension in OPEC, but, paradoxically, it can also work as a stabilizing factor.[20] The reason is that one member country, Saudi Arabia, is predominant and can regulate the supply and price of oil, within certain limits. The future of OPEC and the price of oil will to a large extent depend on the position of Saudi Arabia within the cartel and on Saudi Arabia's preferences and policies.

In the oil market there is of course a relationship between supply policy and price policy. An OPEC decision to set the oil price at a given level is meaningless unless the members manage to set their production at a level of supply that supports the desired price. Therefore, in the case of a price rise, if less is demanded, OPEC as a whole must be able and willing to reduce oil supplies to maintain this price. If demand grows and OPEC wants to hold prices stable, then it must be able and willing to increase supplies accordingly. Thus, when OPEC agrees on a price, it implicitly agrees on a certain supply policy, without making any formal arrangement as to which countries will make the necessary production decisions to realize the price. It is the ability to restrain or to expand production that gives the cartel

---

[19]*Geopolitics of Energy*, p. 141ff.

[20]Dankwart A. Rustow, "U.S.-Saudi Relations and the Oil Crises of the 1980s," *Foreign Affairs*, April 1977, pp. 494–516, and Dankwart A. Rustow and John F. Mugno, *OPEC—Success and Prospects*, New York University Press, New York, 1976, p. 97ff.

its punch, and also gives those members with the freedom to limit supplies the greatest power in determining prices.

In practice, it is Saudi Arabia that plays the decisive role as the regulator of supply, both in OPEC and in the international oil market. Declining oil reserves and levels of production in several other OPEC countries make Saudi Arabia even more important as a key supplier. This will give Saudi Arabia greater leverage, but it also puts more pressure on Saudi Arabia's oil reserves, and this could begin to present a dilemma. How far and under what circumstances is Saudi Arabia able and willing to increase or decrease its oil production in order to defend a given price for oil?

Even though Saudi Arabia's increasing significance as a supplier gives it more leverage within OPEC and on the world oil market, it still does not have complete control. The threat to increase the price of oil is only credible to the OECD countries if Saudi Arabia can afford to decrease output enough to drive up prices and if the demand for oil is sufficiently high. With the other OPEC countries, the threat to flood the market with cheap oil is only credible if Saudi Arabia has enough excess production capacity and if demand for oil is sufficiently low. Also, the possibility of an opponent or a rival to Saudi Arabia within OPEC cannot be excluded. An alliance of OPEC members of the first group, those with large populations and small reserves, and some countries from the second group, those with small populations and large reserves, might be able to afford a cutback of production larger than Saudi Arabia's excess capacity. Thus they might force through at least a short-run price increase against Saudi wishes.

To get a clearer picture of possible conflicts among OPEC members, I propose a rudimentary model of OPEC behavior. It describes how OPEC responds to the demand for oil on the world market. There are two key concepts here: the "residual view" and the "requirement view," which are both based on the distinction between the two OPEC groups.

According to the residual view, OPEC countries with large populations and small reserves produce as much as is needed to satisfy their income requirements. The residual demand for oil is then passed on to the second group of countries, those with large reserves and small populations. Here, production is set according to preferences of conservation and price development, with Saudi Arabia taking up the remaining slack in the system.

The requirement view sees the oil exports of all OPEC countries as governed by specific economic considerations. The countries of the first group, with large populations and small reserves, produce according to their income needs, which may be less than full-capacity production, and they also produce according to some depletion policy. As a result, the production policies of these countries vary. Some, such as Algeria, Iran, and Venezuela, produce at a high rate of depletion, while Indonesia, Iraq, and Nigeria are slightly more conservationist.[21] The countries of the second group, with small populations and large reserves, produce according to a trade-off between short-term income needs and long-run policy goals. Their freedom in choosing their level of production is determined by limits imposed for technical and conservation reasons and by the volume of oil exports needed to satisfy their income requirements. They also take into consideration the long-run demand for OPEC oil.

The countries of the second group can exercise leverage on the price of oil in two ways. Their excess capacity allows them to change prices by increasing output, and, conversely, they can influence prices by their ability to reduce output, which I call the "withholding potential." Excess capacity has in the past contributed to the instability of cartels.[22] Because oil is a finite resource without competitive substitutes, the reduction of output, or the withholding potential, can be as important as excess capacity in determining the market price. The oil price can thus be driven downward by the use of excess capacity because this increases supplies in relation to demand, and the price can be driven upward by the withholding potential because this reduces supplies in relation to demand. In any conflict within OPEC over the price of oil, the relationship between the excess capacity of one coalition of members and the withholding potential of the other coalition will determine the relative strength of the two

[21]Eckbo, *Future of World Oil*, pp. 90–91.
[22]Ibid., p. 49.

groups. This in turn will be the basis for the strategies adopted by each coalition and will probably determine the outcome of the conflict.

The price of oil influences the distribution of output between the two OPEC groups. The countries with large reserves and small populations control the price because they can vary their output. This in turn allows them to influence the amount that the countries of the first group export. For example, if the Saudis increase their exports, this will cause the oil price to fall. In order to maintain the old price, the countries of the first group must cut back their production correspondingly, which in turn limits their income.

We can now begin to get a dynamic picture of OPEC behavior. In OPEC disagreements over prices, certain strategies can be derived for the different types of OPEC countries. For those desiring a higher price for oil, the withholding potential is the key. The ability to reduce production can be used to drive up prices, so it is imperative that they keep their income needs small to maintain this flexibility. The countries of the first group have an obvious interest in high prices because they want to get the most from their small reserves, whereas Kuwait is motivated more by political and ideological interests. For the countries desiring a more moderate price development, mainly the second group of countries with the exception of Kuwait, it is of strategic importance to maintain their excess capacity so that they can expand production and stabilize prices if their rivals reduce output in an effort to raise the price. In a situation of potential conflict it is, paradoxically, of interest to the first group of countries that the second group has high production and little excess capacity, and it is of interest to the second group of countries that the first group has low production and a limited capacity to reduce output.

Some hypothetical examples illustrate how those two coalitions within OPEC might interact under various conditions. If, in a low-demand situation, Saudi Arabia wanted to strengthen its position as the world's leading exporter of oil, it could increase the price of oil. The countries of the first group would then be getting more income and could probably reduce the volume of their oil exports to conserve their limited reserves. Conversely, in a situation of increasing demand, Saudi Arabia might want to reduce the pressure on its oil reserves. It would then decrease the price of oil, hoping that the countries of the first group would increase the volume of their oil exports in order to avoid losing income.

OPEC countries other than Saudi Arabia can also initiate action that will improve their position. Because of its large reserves and small population, Kuwait, on its own, could reduce its oil exports considerably in order to force the price up by creating a shortage. It could also increase its oil exports in order to lower the price. On the other hand, the countries with small reserves and large populations can only afford to moderately reduce their output in any effort to force the price up. They could also try to underbid Saudi Arabia in order to increase their incomes, but because of their limited excess production capacity their position is not strong.[23] Because of this weakness the countries of the first group would have a better chance in an alliance with a surplus country, for example Kuwait. Such a coalition could more easily compete with Saudi Arabia in efforts to drive the price of oil upward.

Saudi Arabia could respond to such an alliance in three ways. (1) It could seek a compromise on the price question. (2) It could resist the increase by maintaining its price without changing the volume of its exports significantly, creating a double market, as was the case during the first six months of 1977. (3) The Saudis could also increase their exports in an effort to keep the oil price stable, but this would place severe pressure on their own reserves.

To sum up, conflicts within OPEC are going to be decided by a combination of external and internal factors, the most important ones being the demand for OPEC oil, excess production capacity, and the income requirements of the various countries. Even if conflicts do not occur openly, the combination of these factors will determine the relations of strength within the cartel.

OPEC stability depends basically upon a convergence of long-term economic and political interests between the two groups.

[23]Rustow and Mugno, *OPEC*, p. 100ff.

In short, a mood of compromise must prevail. Depending on the demand for OPEC oil, Saudi Arabia must be willing to opt for a policy of limited production, if need be, in order to ensure that the countries of the first group get the income that they need. Furthermore, the countries of the first group, possibly together with Kuwait, must be willing to refrain from using their leverage so that the pressure on Saudi oil reserves does not reach unacceptable proportions. Without such restraint there is a possibility that OPEC could collapse through price competition. This could take the form of a double market, or a spiral of underbidding, depending on the conditions.

A spiral of underbidding could occur in a situation of relatively low demand for OPEC oil. Some countries of the first group, possibly allied with Kuwait, might need to expand their revenues. If Saudi Arabia opposes a price rise, these countries could increase production and underbid the Saudi price, attempting to get a larger share of the market and thus higher incomes. However, in response the countries of the second group—in particular, Saudi Arabia—could expand their output and lower the price even more. This spiral of underbidding would ultimately be stopped by the limited excess capacity of the countries of the first group in relation to that of Saudi Arabia. Consequently, the market would stabilize at a point where the countries of the first group produce at their full capacity but with lower oil revenues.[24]

A two-tiered market could develop in a situation of high demand when there is a conflict over the price. Say the countries of the first group, again possibly allied with Kuwait, demand a substantial price increase, but Saudi Arabia refuses. Saudi Arabia threatens to flood the market with oil if the others reduce production. The coalition decides to call Saudi Arabia's bluff and uses its combined withholding potential to reduce output to a level that just satisfies its income requirements. Saudi Arabia then has the choice of (1) using its excess production capacity to increase output and offset the reduction made by the others, (2) complying with the price demands of the others, with a corresponding loss of political standing, or (3) opting for a two-tiered

[24]Ibid., pp. 102–103.

market, perhaps with the help of Qatar and the UAE. In a two-tiered market Saudi Arabia and its allies would go on selling oil at the old price and in the same quantities, with the rest of OPEC and the other oil exporters getting a higher price. The price hike would be limited by the existence of a two-tiered market. This would last as long as Saudi Arabia could not or would not increase its output to drive the price back down. If the Saudis were already producing at a level close to their conservation ceiling, they might accept the two-tiered system rather than draw on their reserves.

A two-tiered OPEC has serious political implications. A refusal by the Saudis to yield to OPEC pressure for higher oil prices will increasingly isolate them within the Middle East.[25] Saudi Arabia would then find itself in a less friendly and a less stable political environment. Such a change could have political repercussions inside Saudi Arabia itself, creating internal splits over official policy. Saudi Arabia would have an interest in joining the new cartel in order to exercise influence in it and to keep relations with its neighbors relatively friendly.

The upper level of the cartel might gain tacit support from other oil exporters, which for political reasons would not want to have a formal link to it. Possible candidates include Egypt, Mexico, Britain, Norway, and the U.S.S.R., if it remains an exporter of oil. A divided cartel could temporarily create short-term commercial dividends for the consumers. However, in the long run it could also isolate the majority of oil exporters from the OECD, seriously weaken the influence of Saudi Arabia in OPEC and in the Middle East, and increase the influence of such oil exporters as Iraq, Libya, and the Soviet Union. Iran would be tempted by its economic interests to participate in the upper level of the cartel, and in this case Western influence in Iran would be weakened, perhaps to the benefit of the Soviet Union. This could imply a sharp reduction in Western influence in the whole of the Middle East.[26]

Finally, a setback in the settlement of the Arab-Israeli conflict

[25]Turner, "North-South Dialogue," p. 60.

[26]U.S. Congress, Senate Committee on Energy and National Resources, *Access to Oil—The United States Relationship with Saudi Arabia and Iran*, 95th Cong., 1st Sess., 1977, p. 111ff.

could have important consequences for oil supplies and OPEC cohesion. In many Arab countries there might be a renewed call to use the oil weapon, and consequently pressure to do so would be exerted on Saudi Arabia. Saudi Arabia would face a choice of risking a conflict with other Arab countries or with the West. A conflict with other Arabs would probably be untenable for Saudi Arabian internal politics and might eventually lead to a more radical anti-Western regime. Another possibility is that in a double market Saudi Arabia might furnish relatively cheap oil to the United States because of their close bilateral relationship, forcing Western Europe and Japan to compete with the rest of the world for more expensive and perhaps less secure non-Saudi oil. This would of course create severe tensions within NATO. Table 19 summarizes a whole range of scenarios, including medium-demand situations.

TABLE 19
OPEC Survival and Collapse at Different Levels of Demand

Demand	Dominant conflict dimensions	Survival	Collapse
Low	Distribution of output	Compromise, first group giving in	Underbidding
Medium	Price Distribution of output	Compromise, Saudi Arabia keeps control, second group giving in	Underbidding, two-tiered market
High	Price	Compromise, second group giving in	Two-tiered market

FIVE

The Political Economy of Oil Prices

This chapter will examine some critical components of oil prices that are both political and economic. It stresses the interaction of OPEC and the OECD area, rather than the position of one particular side. The analysis will start with a look at the interdependence between the two groups and then turn to the influence of competition, the theoretical limits of prices, and the distribution of the oil rent. It shall then attempt to pull together these observations by looking at the future prospects for oil prices.

INTERDEPENDENCE OF CONSUMERS AND PRODUCERS

In the Second Oil Regime, the response of OPEC to the OECD area has taken place within a framework of mutual interdependence. Four different levels can be distinguished:

- There is a mutual dependence based upon oil trade: The two sides now represent respectively more than 90 percent of oil imports and exports.
- There is a mutual dependence based upon trade outside of oil: The OECD countries are the major suppliers of food, consumer goods, capital goods, arms, and modern technology to the OPEC countries, and thus the OPEC countries provide important export markets for the OECD countries.

85

- There is a mutual financial dependence: The OECD countries are dependent on the recycling of OPEC financial surpluses, and several of the most important OPEC countries have increasing financial interests in OECD countries, implying that the economic health of the OECD area determines the return on the OPEC financial investments.

- There is a mutual political dependence created by the situation in the Middle East. This particularly concerns the United States, which has leverage over Israel, and Saudi Arabia, which has some leverage over the other Arab countries.

Within the context of economic interdependence between the OECD and the OPEC countries, the freedom of action of each party can be defined as the degree of independence each has from the other, that is, the ability of each to do without the goods and services provided by the other party. Thus, for the OECD countries, freedom of action is a function of how much they need OPEC oil, need to make exports to OPEC countries, and need the recycling of OPEC money. Correspondingly, for the cartel members, freedom of action is a function of how much they need to sell oil to OECD countries, to import goods from the OECD countries, and to receive income from financial investments in the OECD area.

As we have seen above, both groups can, to a certain extent, regulate their dependence on the other through their economic policies. The relationship between economic policy and energy policy in the OECD countries determines the level of oil imports. Correspondingly, in the OPEC countries, the relationship between economic policy and oil-production policy determines the level of oil exports. Thus economic interdependence between OECD and OPEC countries is complementary and gives an impression of symmetry. However, this relationship is not balanced. The OPEC countries are less dependent on the OECD countries because, as a group, they export oil far beyond their economic needs, maintaining a substantial surplus on their balances of trade and payments. In addition, markets for oil outside the OECD area are growing. OPEC members can thus relatively easily afford to do without much of their income from the OECD countries. Of course, the OPEC countries do require imports of capital goods, consumer goods, and arms from the OECD countries, but a reduction of these imports would not harm the OPEC countries too much. Also, income from investments in the OECD area provides an additional source of revenue to the financially significant OPEC countries. Conversely, the OECD countries are far more dependent on OPEC for oil, export markets, and financial recycling, which gives the OPEC countries the upper hand. Thus it seems that, in any confrontation between the two groups, the OPEC countries would now and for several years to come have the odds on their side.

When we look to the end of the century, the picture is different. For the OPEC countries two parallel processes are at work—the depletion of oil reserves and the growth of income requirements. This means that the OPEC countries must more or less synchronize the development of other sources of income with the depletion of their oil reserves or risk facing a period in the next century with depleted oil, no alternative sources of income, much larger populations, and maybe no alternative sources of energy. The OPEC countries are caught in a race with time, and their superiority in relation to the OECD countries is of limited duration.

Eventually the OECD countries will develop alternative sources of energy, reducing their dependence on OPEC oil, on exports to OPEC countries, and on the recycling of OPEC money. In fact, in the long run the OPEC countries might become dependent on the OECD countries. The OPEC countries with large populations and small oil reserves will probably need the markets of the OECD area for their other exports. The financially significant OPEC countries will eventually need the return on their investments in the OECD area as an alternative source of income. Furthermore, both may need imports from the OECD countries even more than is the case now, given economic and demographic growth.

This prognosis implies that the OPEC countries must exercise their superiority with caution. Questions of oil prices and supplies involve trade-offs between oil and financial interests for the countries with large reserves and small populations, and trade-offs between short-term and long-term interests for all of OPEC. In many ways, these trade-offs are harder on the countries of the second group than on those member countries with large populations and small oil reserves. The needs and choices of the latter are, as we have seen, very well-defined.

For OPEC countries, these trade-offs can be seen as portfolio management questions, that is, as questions of minimizing risks and maximizing returns according to the existing options. The alternatives are essentially (1) investing in domestic economic development, (2) investing in assets abroad, or (3) investing in oil in the ground.[1] Each option has its risk and its potential return. Economic development carries the smallest risk and the largest return for the OPEC countries with large populations and small oil reserves. The OPEC countries of the second group, with small populations and large oil reserves, cannot use this option to the same extent, because of the limited absorptive capacity of their economies, and consequently they should calculate with a low social rate of return. They are therefore faced with a choice of either producing oil and investing in assets abroad or not producing oil and investing in oil reserves at home.[2] By investing in oil reserves at home the country avoids the risks of investment abroad, such as depreciation through inflation, changing exchange rates, losses, or even nationalization. But keeping the oil in the ground does carry the risk that in the future the export value of a barrel of oil will be less than the future value of a properly managed investment bought at an earlier time with the same oil.[3] Consequently, diversification could be the right choice. However, so far the return on foreign investment for most OPEC surplus producers seems to be rather close to zero.

This choice is complicated by the fact that the OPEC countries of the second group, and Saudi Arabia alone, have so much weight in the international oil market and in the world economy that their decisions have an immediate impact on the price of oil

[1]Anwar Jabarti, "The Oil Crisis: A Producer's Dilemma," in Ragaei El Mallakh and Carl McGuire (eds.), *U.S. and World Energy Resources: Prospects and Priorities*, ICEED, Boulder, Colo., 1977, pp. 130–131.

[2]Ibid., p. 130.

[3]Ibid., p. 131.

and the flow of international capital. Here the parallel to the portfolio choices of private investors ends. Instead, OPEC decisions become a question of careful balancing of oil interests with financial interests, and of short-term versus long-term interests. For example, if the OPEC surplus countries, or Saudi Arabia alone, opt for investing much more in oil reserves at home, this would hurt the world economy and compromise their financial interests as well as their overall long-term interests. If, on the other hand, the surplus countries, or Saudi Arabia alone, should decide to expand their investments in the OECD area, this might put severe pressure on their oil reserve, which would also compromise their long-term interests.

In recent years the Saudi position on the oil price has changed back and forth between advocating a freeze and accepting a limited rise in price. This results in part from developments in the Middle East conflict and in North-South relations. An additional factor is the Saudi need to show its muscle in OPEC from time to time. It also reflects the real Saudi dilemma in matters of oil prices and supplies. With growing pressure on its oil reserves, caused by increasing demand and declining production in other OPEC countries, this dilemma can only worsen. Today Saudi Arabia's strength lies in the fact that it has no depletion policy and no fixed ceiling for its oil exports. The present rate of depletion is close to 60 years. With improved recovery and possibly some new discoveries, reserves might be extended. Nevertheless, Saudi Arabia will feel more and more pressure on its reserves, and we can reasonably suppose that sometime during the 1980s Saudi Arabia will have to impose a depletion policy and put a ceiling on production.[4] Realistically this ceiling could be anywhere between present levels of output (about 440 million tons in 1977) and a figure two or three times as high.[5] In any case, the mere existence of a Saudi depletion policy could have a profound impact on oil prices. Thus it could be argued that OPEC currently functions less as a cartel that

[4]Carroll L. Wilson (ed.), *Energy: Global Prospects*, Report of the Workshop on Alternative Energy Strategies, McGraw-Hill, 1977, p. 131ff.
[5]Ibid., p. 135ff.

keeps oil prices artificially high for a limited period of time than as an organization that moderates the development of oil prices both for political reasons and with reference to long-term opportunity costs. This exemplifies the complexity of the interdependence between OPEC and the OECD area.

COMPETITION

Under the Second Oil Regime, the relationship between supply and demand seems to have had a stronger impact than it did in the First Oil Regime. From 1974 to 1977 the real price of oil fell by about 20 percent. Much of this decline was due to the slack in demand and the fact that the spot market for oil has generally shown lower prices than the OPEC price. With the partial elimination of vertical integration through the nationalization of oil production in OPEC countries, there is also a larger number of mutually independent buyers and sellers, and the intermediate market is now potentially much more important. This could create more competition and stronger price fluctuations. Under the Second Oil Regime the spot market and short-term transactions are likely to cover a greater proportion of international oil trade than they did previously. As OPEC countries diversify into downstream operations like refining, transportation, and even marketing, stronger competition might emerge between them, and this could move the prices for oil products and maybe even crude oil downward.[6]

It should also be noted, however, that the decline in the real price of oil took the form of an erosion of the nominal price through inflation, and an unwillingness in OPEC to increase the nominal price of oil between 1974 and 1977. This might have been caused by OPEC apprehensions about the market, but concern over the long-run health of the OECD economies could also have played a role. Since 1974 there has been a greater diver-

[6]Louis Turner and James Bedore, "Saudi and Iranian Petrochemicals and Oil Refining: Trade Warfare in the 1980s," *International Affairs*, October 1977, pp. 572–586.

sification of oil prices according to quality, with particularly low sulphur oil getting a high price.

Over the coming years the real price of oil will be influenced mainly by two sets of factors: the relationship between supply and demand and the degree of competition. If demand is stagnant and downstream involvement by OPEC countries grows, then greater competition and declining real oil prices could well result. This would be a rather short-term phenomenon, as declining real oil prices are likely to foster more active cooperation among OPEC countries to offset the decline in real income. However, with increasing downstream involvement cooperation would become more difficult, and the task of deciding on the distribution of income might have to be resolved by OPEC. This would make the cartel more cumbersome, creating severe strains and perhaps leading to a temporary breakdown. This would not, however, expand the resource base, and, given the limited supply potential, chances are that a cartel would be reestablished after a while. The historical precedent for this is the collapse of the cartel of oil companies in 1931 and its reestablishment in 1934.[7] Thus, within a general trend of long-term price rise, there could be quite substantial short-term fluctuations, both for market reasons and for political reasons. This raises the question of possible limits to the price of oil.

THE PRICE LIMITS

There are two separate sets of limits that apply to the price of oil: One is based on energy production, and the other is defined in international financial terms. Within the energy-production dimension the limits are the cost of production in the most accessible areas and the cost of substitutes. In terms of international finance the limits are defined by the ability of the countries that import oil to pay for it and the income requirements of the countries that produce oil.

[7]Christopher Tugendhat and Adrian Hamilton, *Oil—The Biggest Business*, Eyre Methuen, London, 1975, p. 97ff.

It is a characteristic of the oil market that in terms of energy production the difference between the upper and lower limits is large.[8] The lower limit can be estimated as the costs of production in the Arabian-Persian Gulf area, about $0.15 per barrel, plus transportation, refining, and marketing costs, together perhaps $0.60 a barrel in Western Europe or on the East Coast of the United States. The upper limit is not so sharply defined. Synthetic oil (oil from coal, oil shale, tar sands, heavy oil, etc.), which can immediately replace conventional oil at its end uses, is generally estimated to cost two to three times the present price of oil—in the range of $25 to $40 a barrel.

In evaluating the upper limit two considerations are important. First, the production of synthetic oil is not yet undertaken commercially on a large scale. Many technical and environmental problems are still unresolved or have not yet been confronted, and costs could well escalate even further with larger-scale production. Second, the cost of alternatives has risen with the price of oil. As oil prices moved up in 1973–1974, the prices of coal and uranium rose also, without mining becoming much more difficult and without the resource base being significantly eroded, as was the case with oil. In addition, the cost estimates for new sources of energy also rose substantially. They were ahead of the price of oil before the price rise, and they are further ahead now.

This does not indicate the existence of a large-scale international energy conspiracy, but it does show that oil is a price leader for energy in general. The reason that oil has this role is that supply and demand for other sources of energy are also relatively unresponsive to price changes, and competition is far from perfect. Oil might keep this role as a price leader for a long time. Thus, as the price of conventional oil increases, the prices of other forms of energy may rise as well. Also, an interaction between the price of oil and the price of uranium is a possibility. In recent years the price of uranium has increased much more than the price of oil. If current nuclear programs are pursued and

[8]Dankwart A. Rustow and John F. Mugno, *OPEC—Success and Prospects*, New York University Press, New York, 1976, p. 110.

the development of the breeder reactor is postponed, the demand for uranium could grow enough to push the price up very high. Then uranium might eventually become the price leader for energy, with oil and coal following. This indicates that, theoretically, the price of energy could get very high and that the price mechanism alone is not particularly effective in stimulating the smooth transition from one source of energy to another.

In international financial terms the lower limit for the price of oil is determined by the income requirements of the OPEC countries. For 1975 this lower limit was about $5 a barrel, or less than half the international oil price. At that price the OPEC countries as a group would have had an even balance of trade. The problem is, of course, that income needs vary within the cartel. For some member countries with large populations and small oil reserves, the lower limit has already reached the actual price of oil, and these countries will in the coming years increasingly need a rise in the price of oil. At the same time, OPEC as a group will continue to have a large financial surplus.

In the international financial dimension of oil prices it is difficult to establish precisely what the upper limit is. The price of oil is already a serious burden to the balances of payments of many OECD countries. The price increase of 1973–1974 represented a transfer of income from the OECD countries to the OPEC countries of less than 2.5 percent of OECD-area GNP. It is difficult to assert that this was a serious burden for the OECD countries collectively. But it hit some countries more than others; Italy, Greece, and Denmark are typical examples. Of course, the price of oil is not the only source of economic trouble in these countries. Japan is extremely dependent on imported oil and manages to maintain a positive balance of trade. West Germany is a similar case. Therefore, it is hard to establish the price level of oil that will seriously impede economic growth in the OECD countries. It can be argued, however, that a sudden increase is more harmful than a gradual rise in price over time.

In the international financial dimension, a better picture of the upper limit to the price of oil can be gained by looking at consumer behavior and its effect on producers. A price increase is rational to the producer as long as the price elasticity of demand remains less than 1.0, that is, as long as an increase in the price does not bring about a reduction of demand that causes total income to decrease.[9] Where this point is depends on a multitude of factors, such as the income levels and preferences of consumers and taxes in consumer countries. In any case, this critical point for demand elasticity appears quite distant. Recent experience indicates that the short-term price eleasticity of demand is at most 0.1.[10]

Both dimensions of oil prices—the energy-production and international financial limits—suggest some optimal or correct price for oil, but both systems are flawed. In terms of energy production the correct price of oil is based on the marginal cost of production, that is, the cost to the producer of supplying one more barrel of output. Some believe that this system of pricing provides the optimal use of finite resources and stimulates the development of alternatives, when they become necessary.[11] From this point of view, the price of oil was too high under most of the First Oil Regime because theoretically it ought to approach the cost of production in the most easily accessible areas.[12] However, from this point of view it is also clear that the price of oil has been too low under the Second Oil Regime because it is still insufficient to stimulate the large-scale development of alternative energy sources.[13] This shows that the oil price has been influenced under quite different circumstances by factors other than marginal costs.

In international financial terms it can be argued that the correct price of oil is determined by the needs of the exporting and importing countries, principally by their balances of payments. Such pricing seems to provide an optimal distribution and use of limited economic resources, and can stimulate the economic development of both producers and consumers. According to

[9]M. A. Adelman, "Need for Caution over Prices," *Petroleum Economist*, September 1977, pp. 359–360.

[10]*Exxon, World Energy Outlook*, New York, Exxon, 1977, p. 6.

[11]M. A. Adelman, *The World Petroleum Market*, The Johns Hopkins University Press, Baltimore, 1972, p. 14ff.

[12]Ibid., p. 39ff.

[13]Wilson (ed.), *Energy: Global Prospects*, p. 89ff.

this perspective the price of oil is too high as long as it gives the oil exporters as a group a substantial financial surplus and, on this same basis, the price of oil was too low under the First Oil Regime. In energy-production terms the sudden price rise of oil can be seen as rational, given the historical shift and the reversal of long-term marginal costs. But from the international financial point of view the sudden price rise is seen as undesirable, creating a sudden transfer of income and a distribution of limited economic resources that is less than optimal.

THE CRITICAL PRICE POINT

If allowed to increase, the price of oil will, at some point, make the exploitation of alternative sources of energy economical for both producers and consumers. What this price is and when it will be reached are obviously crucial questions. The combination of the price level and the timing can be called the critical price point (for example, the development of alternative sources of energy requires a price of x dollars per barrel in year y). The critical price point indicates the average increase in the real price of oil that will be required over the coming years to make the transition to other energy sources (see Table 20).

A moderate growth of demand for oil stretches out the period during which conventional oil is relatively abundant, and so diminishes the annual price rises required for a transition to alternative energy. In the same way, limited progress in reducing the costs of alternative sources of energy could increase the growth rate of prices that is required for the transition to these other sources. A combination of two such changes, shifting the critical price both in time and in level, would have a significant effect on the rate of price changes needed. For example, a movement of the critical price point from $25.00 a barrel in 1985 to $22.50 a barrel in 1990 reduces the required price increase from 8.1 percent a year to 4.6 percent a year. Such a move would definitely have an important impact on the relations between producing and consuming countries.

TABLE 20
Growth Rates of the Real Price of Oil According to Different Critical Price Points
(Basis: $11.51 a barrel in 1975. Yearly growth rates required from 1975.)

Year	Price level ($ per barrel)					
	17.50	20.00	22.50	25.00	27.50	30.00
1985	4.3	5.7	6.9	8.1	9.1	10.1
1990	2.8	3.8	4.6	5.3	6.0	6.6
1995	2.1	2.8	3.4	4.0	4.5	4.9
2000	1.7	2.2	2.7	3.2	3.6	3.9
2005	1.4	1.9	2.3	2.6	3.0	3.2
2010	1.2	1.6	1.9	2.2	2.5	2.8
2015	1.1	1.4	1.7	2.0	2.2	2.4

SOURCE: The figures are based on the author's own calculations.

THE OIL RENT

Both the size and the distribution of the oil rent—the sum of profits and incomes from oil—distinguish the Second Oil Regime from the First Oil Regime. Under the First Regime the oil rent was several times the cost of oil, and the bulk of this, probably about 90 percent, went to the companies and governments of consuming countries, while less than 10 percent went to the producing countries.[14] Under the Second Oil Regime, the total oil rent is considerably larger because the price of oil has increased without a corresponding increase in average production costs. The oil rent has also been redistributed, to the advantage of the OPEC countries. They now get about a third of the oil rent, and the relative share of oil companies and OECD governments has decreased. However, because of the price increase,

[14]Ali M. Jaidah, "Pricing of Oil: Role of Controlling Power," *OPEC Review*, June 1977.

TABLE 21
Distribution of the Rent from Oil

	$ per barrel Percentage			
	1961	*1973*	*1975*	*Change 1975/1973*
Revenue of producing country	$0.76/6%	$2.30/11%	$10.10/34%	+ $7.80 + 339
Taxes by consuming country in Europe	$7.10/52	$11.40/56	$14.90/45	+ $3.50 + 31
Company margins and cost elements	$5.70/42	$6.80/33	$8.20/25	+ $1.40 + 21
Total weighted average	$13.60	$20.50	$33.20	+ $12.70 + 62

SOURCE: Ali M. Jaidah, "Pricing of Oil: Role of Controlling Power," *OPEC Review*, June 1977, p. 15.

the actual revenues of the oil companies and OECD consumer governments from imported oil are greater than they were under the First Oil Regime. This means that the OPEC price rise not only has been passed on to the consumer but has been further increased on the way, in order to allow for larger company margins and government revenues.

Clearly, control over the final price of oil—the amount paid by individual consumers—is not exclusively in the hands of the oil-producing countries. Instead it is shared among OPEC, the oil companies, and the governments of consuming countries.[15] Both oil companies and consuming country governments thus still reap large profits from oil. This is shown by a comparative survey of the composition of final oil prices (Table 21).

The case of France (Table 22), where local taxes on oil products are relatively high, illustrates this development clearly.

Because of their contribution to final oil prices, governments

[15]Hendrik A. Houthakker, *The Price of World Oil*, The American Enterprise Institute of Public Policy, Washington, D.C., 1975, p. 33.

TABLE 22
Size and Distribution of the Oil Rent in France

	French francs per metric ton		
	1970	1973	1974
Average retail price	470.00	520.00	820.00
Average import price	56.00	81.50	299.00
Average tax plus cost	46.00	64.00	268.00
Costs of production	8.00	6.00	5.00
Costs of transportation	10.00	17.50	31.00
Costs of refining	25.00	25.00	25.00
Costs of distribution	35.00	35.00	35.00
Average total costs	78.00	83.50	97.00
Average oil rent	392.00	436.50	723.00
Oil rent/retail price	83.4%	83.9%	88.2%
French government revenue	265.00	262.70	315.00
Company profits	89.00	115.80	145.00
Producing country revenue	38.00	58.00	263.00
Percentage of oil rent:			
French government	67.6%	60.2%	43.6%
Oil companies	22.7%	26.5%	20.1%
Producing countries	9.7%	13.3%	36.4%

SOURCE: Jean-Marie Chevalier, *Le Nouvel Enjeu Pétrolier*, p. 14ff.

in the consuming countries can influence the impact of any increases in the price of oil. In particular, they can choose, to a certain extent, between protecting their national economies from the price increase and magnifying its domestic impact. This is illustrated by a look at the experience of the OECD area.

For the OECD countries, the increase in the international oil price in 1973–1974 had two distinct effects. First, it created a new burden on the balance of payments. Second, consumers reduced oil consumption only slightly, so they spent less on other goods and services, thus damping the national level of economic activity. Because of the wide range of uses for oil in numerous products and processes, the prices of many goods and services rose, adding to inflationary pressures. Moreover, given the economic situation in 1973–1974 of inflation and restrictive economic policy measures aimed at checking inflation, the oil price increase only aggravated these conditions. It thus seems that the policy responses of many OECD governments to the price increase exacerbated both inflation and recession. Economic policy was perhaps more responsible for the recession than the rise in the international oil price. The response of many OECD governments, higher taxes on oil products, probably only marginally contributed to a reduction in oil consumption and limited only slightly the pressure on the balance of payments. This policy actually magnified the impact of the price increase by further neutralizing demand for other goods and services and adding to inflation.

In the face of future oil price hikes, governments have a choice of absorbing the shock either domestically or internationally. If they increase taxes and duties, they will slow consumption and reduce their oil import bill, which limits their vulnerability to OPEC at the expense of the domestic economy. The government's increased taxes and duties are also inflationary and stunt growth at home. Conversely, a decrease in taxes and duties can maintain an almost constant price to the consumer, which insulates the domestic economy, eliminating the inflationary pressure of the oil price hike and avoiding adverse effects on economic growth. However, internationally, pressure on the balance of payments will increase because consumption will stay at past levels and the oil price will be higher.

Recent OECD experience indicates that it is difficult to reduce oil consumption through price changes alone. However, oil consumption is affected by changes in the level of economic activity. Therefore, a recession is one way to moderate oil imports and intensified inflation can reduce the import bill in real terms. But these means seem to be socially expensive, particularly considering their rather low effectiveness. An increasing burden on the balance of payments is tolerated more easily by an expanding economy than by an economy in stagnation or recession. This is especially true if there are effective international institutions to recycle the financial surpluses arising from oil. Thus recent experience seems to argue for a decline in the OECD proportion of the oil rent in the interest of domestic economic health.

The minimum support price of the IEA can be seen as an attempt to overcome OECD internal disagreement and at the same time to retrieve part of the oil rent from OPEC.[16] The present minimum support price ($7 a barrel) has little practical importance, but the principle merits attention. In order to be an effective stimulant to the development of alternative energy sources, and thus add to the energy independence of the OECD area, a minimum support price of perhaps $25 a barrel is needed. This level could be reached gradually over a given number of years to make it politically more acceptable. It could take the form of a price guarantee or subsidy to energy producers and investors in the IEA area. In order to prevent the actual price of oil on the world market from rising to this level, it could be composed in part of a duty on oil imported into the IEA area.[17] In reality this would be a new tax on oil consumers, and the amounts paid in this duty could be channeled to the treasuries of the IEA countries.

Theoretically, such an import duty could prevent the price of oil on the world market from increasing further and might even lead to a drop in the international oil price. This could happen if demand was effectively reduced by the price rise. An important consequence would then be a stronger balance-of-payments position for the consuming countries.[18] Theoretically, this would transfer income from OPEC countries to the OECD area, without alienating either the OECD producers—Canada, Norway, Britain, and, to a certain extent, the United States—or the consuming countries of the industrial core. If such a strategy succeeded, it would seriously weaken or even break up OPEC. A drop in the demand for oil on the world market would induce some OPEC countries to cut output in order to defend the price, while others

[16]Ibid., p. 34.
[17]Ibid., p. 33.
[18]Ibid., p. 32

would increase production in order to maintain incomes. Thus the cohesion of OPEC would be seriously weakened.

In practice, however, such a strategy would probably be seen as a threat by the OPEC countries and would provoke retaliations. The OPEC countries could together, or one by one, reduce their output of oil, and thus create a shortage on the world market with further price increases as a consequence. Also, the OPEC countries could offer favorable bilateral deals to countries refusing to join the IEA plan, or restrict exports to countries that do join. Moreover, there are risks in this strategy: It could provoke political changes in important OPEC countries, leading to policies that would be less friendly to the consuming countries. A final snag is that the minimum price plan requires a long lead time before it could provide IEA countries with a higher level of energy independence. In the meantime, the ability of IEA countries to get oil supplies at prices they can afford would depend on the good will of—or internal strife in—the OPEC countries. A plan to get a larger proportion of the oil rent would only make sense as long as the OECD countries are highly dependent on OPEC oil, but in such a situation its success is dubious.

PRICE PROSPECTS

A low level of oil imports into the OECD area will not silence the demands within OPEC for price increases, as long as the oil price remains below the cost of alternative energy sources. As mentioned earlier, many OPEC countries with large populations and small oil reserves are likely to have trade balance deficits by the early 1980s. They will then have a choice of reducing their domestic programs of social and economic development, increasing their oil exports, or pressing for a hike in the oil price. The last option is the most acceptable to them because their ability to increase exports is limited and a loss of income is obviously undesirable.

An increasing demand for oil in the OECD area, even at a relatively slow pace, will cause the ceilings of production set by many OPEC countries to be reached by the early 1980s. This could happen at the same time as some OPEC members develop negative trade balances.[19] These countries will then have a choice between raising their production ceilings or increasing oil prices. It is not unreasonable to assume that both may take place. An increase in the price of oil would to a certain extent satisfy the members of the cartel that need income or political leverage. An increase in the output of oil would prevent the oil price from rising to levels that are unacceptable to members that still have surpluses and are more politically conservative, especially Saudi Arabia.

The nature of such a compromise will largely be determined by OECD demand for oil. Increasing OECD imports will probably strengthen the position of the OPEC countries wanting significant price increases. But no one knows the level at which increasing OECD demand would eventually destroy the basis for an OPEC compromise on the price question. At a high level of demand for oil, a coalition between members with inadequate incomes and their wealthy sympathizers, perhaps Kuwait, could impose substantial price increases. Saudi Arabia would then have the choice of either accepting a price it finds incompatible with its own long-term market interests or trying to flood the market in order to bring the price down again.

Unless the basic structure of the oil market is transformed or the cost of alternative energy falls considerably, the real price of oil is likely to increase in the medium-term future.[20] Exactly how fast and how much depend on political relations between OECD and OPEC countries, and their own internal cohesion, the development of alternative energy sources, and the economic needs of each side. A clearer picture of the future is needed as a basis for an international energy agreement. We now turn from the analysis of the world oil market to a much closer examination of how it is likely to develop in the 1980s.

[19]J. Alexander Caldwell, "Trends in OPEC Surplus," *The American Banker*, September 22, 1977.
[20]Houthakker, *Price of World Oil*, p. 25ff.

SIX

Problems and Risks in the Second Oil Regime

THE BASIC PROBLEMS

The Second Oil Regime suffers from two inherent faults:

- It fails to reconcile the oil interests of the major partners—the OECD countries and the OPEC countries—and, consequently, has a substantial potential for conflict.
- It fails to secure the oil interests of either side and, consequently, lacks a basis for mutual trust and stability.

The combination of these two flaws, the inability to reconcile and the inability to secure crucial interests for either side, means that the Second Oil Regime is engendering tensions and conflicts that it is not likely to resolve. The Second Oil Regime is as a result less stable than the first. The First Regime did not reconcile the oil interests of consumers and producers, but it did secure the oil interests of the industrialized consumers well enough to provide a solid power base and achieve both relative stability and remarkable longevity, in spite of growing tensions on the world oil market.

At present, the dominance of the United States and Saudi Arabia on either side of the market creates a system of "oil superpower" dominance that, in the short run, stabilizes the Second Oil Regime. Today the Second Oil Regime succeeds, to a considerable extent, in reconciling the interests of the United

States and Saudi Arabia. The United States' desire for greater supplies on the world oil market at moderately increasing prices and the Saudi interest in a stable world economy and a rather moderate rise in the price of oil are complementary. In this sense, the United States–Saudi bilateral link provides the source of stability in the Second Oil Regime. However, this apparent solidity implies a rather static situation on the world oil market and in Saudi Arabia, neither of which is likely in the longer run.[1]

Given the process of rapid economic and social change in Saudi Arabia, it is not realistic to assume that its oil policy will remain unchanged in the medium-term future. Already the Saudi political elite faces dilemmas on oil-depletion policy and on the priority of their ties with the United States relative to links with other Middle Eastern and OPEC countries. In addition, the close bilateral relationship with the United States increasingly isolates Saudi Arabia in the Middle East and in OPEC.[2] In the short run this does not necessarily have any particular political significance, but eventually any Saudi government is likely to be affected by a less and less friendly environment.

Any deterioration in the general political situation in the Middle East could magnify the potential for a change of oil policy in Saudi Arabia. To a considerable extent, the bilateral link between the United States and Saudi Arabia is based on politics, especially on American political leverage with Israel. Any setback of United States diplomacy concerning Israel could provoke substantial political repercussions within Saudi Arabia, possibly changing their policy and destabilizing the Second Oil Regime.[3]

The bilateral link between the Saudis and the United States can also be eroded by the dynamics of the world oil market. Increasing demand for oil in the OECD countries, especially the United States, puts a larger export burden on Saudi Arabia and

[1] Louis Turner, "Oil and the North-South Dialogue," *The World Today*, February 1977, p. 59ff.

[2] Ibid., p. 60.

[3] U.S. Congress, Senate, Committee on Energy and National Resources, *Access to Oil—The United States Relationship with Saudi Arabia and Iran*, 95th Cong., 1st Sess., 1977, p. 60.

plays into the hands of other OPEC countries. In addition, regardless of the demand situation, the increasing income requirements of other OPEC countries also put pressure on Saudi Arabia. Thus, United States–Saudi relations in the context of Saudi domestic politics and Arab and OPEC concerns create a situation in which pressures build up on Saudi Arabia. These come from the Western consuming countries—in particular the United States—that desire more oil at relatively stable prices, and from OPEC and Middle Eastern countries that desire an oil price rise and use of the political leverage derived from oil. This political dilemma within Saudi Arabia increases the potential for a change of policy and the deterioration of United States–Saudi relations. In this way the dynamics of the world oil market undermine the stability of the Second Oil Regime. In addition to this erosion of the institutional framework, there are also more practical sources of instability.

An essential function of the close bilateral relationship between the United States and Saudi Arabia is to stimulate the latter to provide increasing quantities of oil for the world oil market at moderately rising prices.[4] In the short run this has a positive effect for the world economy as a whole and in particular for the industrialized consumers. But eventually this will have significant negative effects because the Second Oil Regime actually aggravates a number of serious structural problems existing in the world oil market. Four of them are definite potential sources of instability in the current oil regime.

First, the discrepancy between the technological horizon and the market horizon of alternative sources of energy is increasing because of the combination of growing pressure on low-cost petroleum and escalating costs of alternative energy. The effort to develop alternative energy sources is frustrated by a lack of political determination and the difficulty of mobilizing the necessary capital. Clearly, the necessary political will and capital flows cannot be mobilized by increasing oil supplies at moderately rising prices. This lack of motivation prepares the ground for a serious discontinuity in energy supply when conventional

[4]Turner, "North-South Dialogue," p. 58ff.

oil production starts to decline and a transition to other sources becomes more urgent. This possibly could provoke a sharp price rise at some point in the near future.

Second, the discrepancy between the demand for oil from oil-producing countries that can expand supplies and their economic need to do so is growing. Supply increases from the few countries with large oil reserves and small populations are an essential factor in the present stable functioning of the Second Oil Regime. Eventually this could lead to an oil supply crisis, if these countries decide to impose oil-production limits more compatible with their own income requirements.

Third, the discrepancy between the income requirements of the oil-producing countries that are unable to expand production and their actual incomes could be a serious problem in the 1980s. Their need for income can only be satisfied if the price of oil stays at a level that can pay for all of their imports. This gives them a bias in favor of high prices. Failure to achieve income needs could lead to domestic political instability in the countries concerned and political strife among the oil exporters, possibly jeopardizing the stability of supplies or the future of the cartel.

Fourth, there is a growing burden on the balance of payments of many industrialized consumers and most LDCs because of their increasing imports of oil. Conversely, there is also the accumulation of large currency surpluses in some oil-producing countries. This structural problem in monetary relations would be aggravated by increasing oil prices, and its continuation prepares the ground for uneven or stop-and-go growth in the world economy, and possibly prolonged economic stagnation.

What we have, then, is a structurally deficient oil regime. Its short-term stability depends on the ties between the United States and Saudi Arabia. This bilateral relationship is influenced by the Arab-Israeli conflict, by North-South relations, and by the internal politics of OPEC. In practice, for the OECD area, oil supplies and oil prices are increasingly dependent upon a bilateral relationship with a single country. This is hardly a stable or desirable situation, even in the short run.[5] What is worse, this

[5]Ibid., p. 60.

short-term stability aggravates the potential for the long-term instability of oil supplies and prices. The longer the stable situation of increasing supplies at moderately rising prices endures, the worse the basic structural problems become. The dependence of the OECD countries on imported oil will continue and most likely increase, and the effort to develop alternative sources of energy will not be further stimulated. In the meantime contradictions within OPEC will grow, and the international monetary system will be further weakened.

There are hopes in some OECD countries that a settlement in the Middle East would permit Saudi Arabia to open the valves, expand production, and exercise price discipline in OPEC. This is a shortsighted solution that in the long run would be detrimental to the interests of the OECD countries. It would make the transition to alternative sources of energy even more problematic. Likewise, competiton within OPEC and possibly a temporary breakdown of the cartel might in the long run harm the interests of the OECD countries and the world economy. It would temporarily create the impression in the West that the importance of the energy problems had been exaggerated, but it would not create more oil, and, consequently, it would make the eventual crunch much worse.

A dynamic picture of the situation is provided by the three scenarios that follow. The scenarios are based on different assumptions about the rate of economic growth and the energy policy of the OECD area. They are intended to illustrate the main options open to the key participants. The construction of scenarios for the interaction between the OECD and the OPEC countries raises a number of problems of method. The scenarios must to a considerable extent be based on assumptions that cannot be proven empirically. Thus the reasons for making particular assumptions are of critical importance, as the assumptions to a large extent determine the outcome. Four sets of problems are of particular relevance.

First, the major question concerning the OECD countries is the relationship between changes in economic growth and the level of oil imports in given years. In this study explicit assumptions must be made on the energy coefficient and the growth of domestic energy production because at this time the full impact of a price rise upon these factors remains obscure.

Second, the role of those outside OECD and OPEC can be of crucial importance in the coming years. This concerns the non-OPEC LDCs, Israel, South Africa, the Soviet Union, Eastern Europe, and China. The OECD's *World Energy Outlook* assumes that the non-OPEC developing countries collectively will become a modest net importer of oil by 1985. This is based on an estimate of almost 9 percent a year as the potential for increasing local energy production, implying a substantial increase in the output of coal, oil, natural gas, and electricity.[6] For example, the output of primary electricity is assumed to grow at a yearly rate of 13.5 percent, and the output of natural gas is assumed to grow at a rate of 10.1 percent. These estimates seem optimistic. However, the estimated yearly growth of energy consumption seems more reasonable, 5.3 percent, given a demographic growth of 2 to 3 percent a year.[7] With less optimistic estimates on LDC energy production, the non-OPEC LDCs collectively remain net importers of oil, as do South Africa, Israel, Cyprus, and a number of smaller countries and territories not included in any other grouping.[8] The future of Soviet oil production and Soviet oil exports is enigmatic. We will assume here that Eastern Europe's oil imports will be balanced by Soviet and Chinese oil exports until 1990. But this assumption is easy to criticize, given the different estimates of Soviet oil.

Third, the growth of domestic oil consumption in OPEC countries could in some cases considerably reduce oil exports. We assume here that oil consumption in the OPEC countries will increase by 8 percent a year. As for energy consumption outside the OECD area, it will be estimated to increase at an average yearly rate of 6 percent, reflecting a rate of economic growth of 5 percent a year and an energy coefficient of 1.2 owing to relatively early phases of industrialization. Energy production outside the OECD and the OPEC areas is assumed to increase at an average rate of 6 percent a year.

[6] *World Energy Outlook*, OECD, Paris, 1977, p. 12ff.
[7] Ibid., p. 82ff.
[8] Ibid., p. 13.

Finally, explicit assumptions must be made about the imports of the OPEC countries and their financial surpluses at different oil prices. The absorption of the OPEC surpluses has been a smaller problem than was once anticipated. This is partly due to regional recycling, that is, economic aid from the surplus countries to other countries in the Middle East.[9] However, a distinction should be made between absorption at a regional level and each country's income requirements because in a situation of crisis or potential confrontation foreign aid can be temporarily suspended more easily than imports can be cut back. In a confrontation the ability to increase or to reduce oil production is also essential for the leverage of each OPEC country or coalition of countries. A country's flexibility is also influenced by the size of alternative sources of income, either exports other than oil or income from investments. In the wake of the oil revolution, imports to most OPEC countries increased at high yearly rates. The U.S. Department of the Treasury has estimated that imports to OPEC countries will continue to grow at a rate of perhaps 15 percent a year at least until 1985. But there are now signs that imports could grow at a slower rate.[10] It is obviously impossible to be certain about these factors, and thus, rough assumptions must be made.

SCENARIO ONE: LOW OECD ECONOMIC GROWTH

Demand for OPEC Oil

Partly because of inflation, balance-of-trade problems, and insecurity about future energy supplies, the rate of economic growth remains low in the OECD countries. It averages 3 percent a year in North America and Western Europe and 5 percent a year in Japan. The energy coefficient before 1980 averages 1.0 in all of the OECD area; after 1980 it stabilizes around 0.8 in

[9]J. Alexander Caldwell, "Trends in OPEC Surplus," *The American Banker*, September 22, 1977, p. 1.

[10]Ragaei El Mallakh, Mihssen Kadhim, and Barry Poulson, *Capital Investment in the Middle East*, Praeger Publishers, New York, 1977, p. 45.

North America and around 0.9 in Western Europe and Japan. The growth of domestic energy production is low in North America (2.75 percent a year) and higher in Western Europe (5.5 percent a year). In Japan the consumption of energy from domestic sources, mainly nuclear, and from nonoil energy imports, grows at an annual rate of 7 percent. In the Third World economic growth is higher. The noncommunist countries outside the OECD and OPEC—called the "rest" in these scenarios—have an average economic growth rate of 5 percent, and an energy coefficient of 1.2 because of the early phase of industrialization. Their domestic output of energy collectively grows at a rate of 6 percent a year. The picture thus given is represented in Table 23.

OPEC Response

The low growth of the OECD economies forces OPEC countries to limit their import growth after 1980. Until 1980 the growth of imports in real terms is 10 percent a year in the countries with large populations and small reserves and 20 percent a year in the countries with small populations and large reserves. After 1980 the import growth rates are halved (5 percent a year in the first group, 10 percent a year in the second group). Other exports grow at a rate of 10 percent a year. By 1980 this creates negative trade balances for almost all OPEC countries, the most notable exceptions being Kuwait and Saudi Arabia. Almost all of the OPEC trade deficit countries produce at full capacity, and domestic oil consumption grows at an annual rate of 8 percent. Both of these factors severely limit the future export potential of several countries.

Price Politics

Because most of the OPEC countries in the first group have negative trade balances, their exports do not cover their income requirements and they have limited political weight in determining prices. Therefore, the price of oil can be assumed to remain relatively stable until at least 1985. After 1985 there is increasing pressure on Saudi Arabia, through a combination of growing demand and a reduction of output in some of the countries of

TABLE 23
Scenario One: Energy Consumption and Production in the 1980s and 1990s, in Millions of Metric Tons of Oil or the Equivalent

	1975	1980	1985	1990	1995	2000
Total Energy Consumption						
North America	1,804	2,183	2,458	2,767	3,116	3,508
Western Europe	1,138	1,315	1,503	1,717	1,962	2,241
Japan	329	418	521	649	809	1,008
OECD Total*	3,361	3,916	4,481	5,133	5,886	6,757
Rest†	752	1,006	1,347	1,802	2,412	3,227
World‡	4,113	4,923	5,828	6,935	8,298	9,984
Domestic Output of Energy						
North America	1,559	1,831	2,097	2,402	2,751	3,151
Western Europe	521	681	890	1,163	1,520	1,987
Japan§	89	125	175	246	344	483
OECD total*	2,209	2,637	3,162	3,811	4,616	5,620
Rest†	499	668	894	1,196	1,600	2,142
World‡	2,708	3,305	4,056	5,007	6,216	7,762
Oil Imports						
North America	295	352	360	365	365	357
Western Europe	617	634	613	554	441	254
Japan	240	293	346	403	464	525
OECD total*	1,152	1,279	1,319	1,322	1,270	1,136
Rest†	253	339	453	606	811	1,086
World‡	1,405	1,618	1,772	1,929	2,082	2,222

*Excluding Australia and New Zealand.
†Including Australia and New Zealand.
‡Excluding the Soviet Union, Eastern Europe, and China.
§Including consumption of imported energy other than oil.
SOURCE: Author's projection.

TABLE 24

Scenario One: OPEC Oil Exports in the 1980s and 1990s, in Millions of Metric Tons of Oil

	1975	1980	1985	1990	1995	2000
The Distribution of						
Oil Exports						
Group I	651	857	808	681	516	294
Group II*	754	761	954	1,248	1,566	1,928
Saudi Arabia	544	537	741	1,027	1,348	1,715
Saudi Excess Capacity						
at Different						
Production Ceilings						
750	193	201	−9†	−304	−638	−1,023
1,000	433	451	241	−54	−388	−773
1,250	693	701	491	196	−138	−523

*Includes Saudi Arabia.
†Negative values indicate the extent to which the demand for exports exceeds the production ceiling.
SOURCE: Author's projection.

the first group. Also, the countries of the first group have significant current account deficits and accumulate large foreign debts. Consequently, they press for a substantial increase in the price of oil, which would improve their situation. A coalition of countries, including Kuwait, threatens a cutback of supplies that might be greater than Saudi Arabia's excess capacity, if Saudi production limits are not set at high levels, for example, at a billion metric tons a year. As a result, sometime between 1985 and 1990 Saudi Arabia yields to the price pressure and the oil price is increased by perhaps 50 to 100 percent over a short time. This helps to improve the financial situation of most other OPEC countries but is detrimental to the world economy.

Political Tensions

The constant price of oil, in real terms, creates tension within OPEC. The countries with large populations and small reserves reach their ceilings of output around 1980, with the exception of Iraq, making them borrowers on the international financial market. Lending to these countries is encouraged, both by Saudi Arabia and by the West, to appease their demands for a higher oil price. However, this is increasingly resented by these countries for political reasons and because the service of the debts becomes a burden. Saudi Arabia is increasingly seen as being closely committed to the cause of the OECD countries, and particularly as acting hand in glove with the United States. Consequently, Saudi Arabia becomes isolated in OPEC, in the Middle East, and in OAPEC.

From the West, Saudi Arabia gets only small rewards. Tension in the Middle East conflict is reduced, but no viable solution to the Palestinian problem is found. In North-South relations only limited progress is made, as the OECD countries claim that their low rates of economic growth prevent considerable concessions to the Third World. In addition, the IEA, under United States leadership, tries to keep member countries together by closer cooperation on the development of alternative sources of energy, offering economic relief to some of the most adversely affected countries. The IEA also tries to prevent price increases by presenting OPEC with a common front and threatening retaliations in the form of trade restrictions.

The low rate of economic growth, particularly in North America and Western Europe, creates unemployment, social tensions, and political instability. Increasingly, economic policies and the priority given to keeping oil imports low are questioned. Some countries adopt protectionist measures to insulate their economies. The LDCs maintain relatively high rates of economic growth, around 5 percent a year, which are facilitated by the constant real price of oil. The poorest LDCs accumulate substantial foreign debts.

In the OPEC countries the process of economic and social change is continued, but those countries with large populations and small reserves, where development efforts are most significant, run into financial and economic problems because of their balance-of-trade deficits. This threatens their political stability. In addition, all OPEC countries are adversely affected by the low rate of economic growth of the OECD area. The countries of the first group find few outlets for their new industrial exports, and their nonoil exports grow at a slow pace. Thus they must consider limiting their imports, even though this will have adverse effects upon their economic development. The countries of the second group continue to accumulate financial assets, but the return on their investments in the West is low partly because of low rates of economic growth there.

When the price shock hits, after 1985, both the OECD countries and the LDCs are particularly vulnerable. This time the OECD countries are confronted with a drastic rise in the price of oil after a prolonged period of economic stagnation, not as in 1973–1974 after a period of sustained growth. Therefore, the inflationary and the recessionary effects are worse. Also, the OECD countries have not been able or willing to finance a strong effort to develop alternative sources of energy because of their low rates of growth. Furthermore, this effort has been inhibited by the stagnant real price of oil. The LDCs are also adversely affected because they are now much more dependent on imported oil for development, and in many cases their current account deficits reach alarming proportions. This adds to international financial instability. According to this scenario, prospects are that the new price rise in the late 1980s will severely damage a world that is still highly dependent on oil, leading to prolonged economic stagnation as well as financial and political instability.

SCENARIO TWO: HIGH OECD ECONOMIC GROWTH

Demand for OPEC Oil

Economic growth picks up in the OECD countries. It averages 4 percent a year in North America and Western Europe and 6 percent a year in Japan. The energy coefficient is reduced, averaging 0.8 in North America and 0.9 in Western Europe and Japan. The growth of domestic energy production is the same as in the previous scenario: 2.75 percent a year in North America, 5.5 percent a year in Western Europe. In Japan energy consumption from domestic sources and nonoil imports grows at 7 percent a year. The rest have the same performance, 5 percent economic growth, an energy coefficient of 1.2 and an average growth rate of domestic energy output of 6 percent a year.

TABLE 25
Scenario Two: Energy Production and Consumption in the 1980s and 1990s, in Millions of Metric Tons of Oil or the Equivalent

	1975	*1980*	*1985*	*1990*	*1995*	*2000*
Total Energy Demand						
North America	1,894	2,287	2,677	3,133	3,668	4,293
Western Europe	1,138	1,379	1,646	1,964	2,344	2,798
Japan	329	438	569	741	964	1,253
OECD total*	3,361	4,104	4,892	5,838	6,976	8,344
Rest†	752	1,006	1,347	1,802	2,412	3,227
World‡	4,113	5,110	6,239	7,641	9,387	11,572
Domestic Output of Energy						
North America	1,599	1,831	2,097	2,402	2,751	3,151
Western Europe	521	681	890	1,163	1,520	1,987
Japan§	89	125	175	246	344	483
OECD total*	2,209	2,637	3,162	3,811	4,616	5,620
Rest†	499	668	894	1,196	1,600	2,142
World‡	2,708	3,305	4,056	5,007	6,216	7,762
Oil Imports						
North America	295	455	579	731	917	1,143
Western Europe	617	698	756	801	824	811
Japan	240	313	394	495	619	770
OECD total*	1,152	1,467	1,730	2,028	2,360	2,724
Rest†	253	339	453	606	811	1,086
World‡	1,405	1,805	2,183	2,634	3,171	3,810

*Excluding Australia and New Zealand.
†Including Australia and New Zealand.
‡Excluding the Soviet Union, Eastern Europe, and China.
§Including consumption of imported energy other than oil.
SOURCE: Author's projection.

OPEC Response

The higher growth rates of the OECD economies make the OPEC countries more confident in their economic policies, and they maintain fairly high rates of imports growth. Imports continue to grow in real terms at 10 percent a year in the countries of the first group and at 15 percent a year in the countries of the second group. Exports other than oil grow at a rate of 10 percent a year. This makes all OPEC countries, excluding Saudi Arabia and Kuwait, have negative trade balances by 1980. Domestic oil consumption increases at an average of 8 percent a year in all OPEC countries.

Price Politics

The increasing demand for oil strengthens the position of the price radicals and, even before 1985, presents Saudi Arabia with a serious dilemma: whether to establish a depletion policy and let prices go up or to try to moderate prices and allow pressure to build on its oil resources. The economic growth of the OECD countries and their sizable contribution to incremental demand lend support to demands among OPEC members for a price rise. The real price of oil remains constant until 1980. Soon after 1980 there is a major confrontation on oil prices in OPEC. An important element is that Kuwait is having a negative balance of trade. Before its financial resources are eroded by deficits, Kuwait prefers to use them as leverage, and offers to finance output reductions in other OPEC countries. Saudi Arabia yields to the increase rather than strain its oil resources further. There is an OPEC agreement on a price escalation, and over a period of five years the real price of oil is doubled.

Political Tensions

The rising price of oil has an adverse effect on the trade balances of most OECD countries and LDCs. There is also a growing concern over the size and availability of oil supplies and a sense of increased vulnerability and dependence. The IEA, under United States leadership, tries to keep member countries together

TABLE 26

Scenario Two: OPEC Oil Exports in the 1980s and 1990s, in Millions of Metric Tons

	1975	1980	1985	1990	1995	2000
The Distribution of Oil Exports						
Group I	651	863	819	681	516	311
Group II*	754	942	1,364	1,953	2,655	3,499
Saudi Arabia*	544	718	1,141	1,732	2,437	3,286
Saudi Excess Capacity at Different Production Ceilings						
750	197	20	−409†	−759	−1,727	−2,594
1,000	447	270	−159	−509	−1,477	−2,344
1,250	697	520	91	−259	−1,227	−2,094

*Includes Saudi Arabia.

†Negative values indicate the extent to which export demand exceeds the Saudi production ceilings.

SOURCE: Author's projection.

by closer cooperation on the development of alternative energy sources and helps finance the effort in the most adversely affected countries. The IEA also tries to prevent further price increases by coordinating diplomatic pressure and presenting OPEC with a united front. But the IEA is increasingly split between countries wanting concerted action against OPEC and countries tempted to make bilateral deals with cartel members to secure supplies and compensate for the high price.

In the OECD countries the rising price of oil accelerates inflation, and increasing current account deficits limit the freedom of most governments in economic policy. Stagflation is the likely outcome. Among the OECD members a breakdown of free-trade policies might even develop as a last resort to protect national interests at the expense of international economic growth. This economic stagnation augments social and political tensions within many OECD countries, leading to increasing political instability.

The high price of oil forces most LDCs to run very high current account deficits. Rapid change in some LDCs brings more radical groups to power, demanding compensation from both OECD and OPEC countries for the high oil price. The higher oil price places obvious strains on Southern unity. Some of the most desperate LDCs threaten to go bankrupt in order to clear their debts. Others make special agreements with the Soviet Union in order to get economic assistance and perhaps oil at a reduced price. Finally, the high price of oil induces wealthier LDCs to develop nuclear power, causing a spread of nuclear military technology.

In OPEC there is an increasing split between Saudi Arabia and the other member countries, with the Saudis being more and more isolated over price questions. For most members the principal effect is huge financial wealth and an increasing political weight in international relations. Domestically, the pressure on oil reserves and the process of economic and social change lead to increasing political instability. Like the OPEC members, the Soviet Union and, to a smaller extent, China gain more international influence through the rising price of oil.

Thus the high demand situation hits the consumers hard. Not only is the economic health of the OECD area endangered but many LDCs are in serious trouble, which undermines Southern unity. The international financial position of the importers is so weak that economic growth levels are not likely to be maintained and recession is probable. The OPEC position is basically one of strength, but there are definite possibilities of OPEC splits over Israel and the Third World, in addition to the political polarization of Saudi Arabia. It is hard to say at what point these tensions will change the oil regime or how much OPEC will exploit its greater political leverage. Certainly the close bilateral ties between Saudi Arabia and the United States will be jeopardized. Within OPEC there is a possibility that a two-tiered cartel will develop if the countries of the first group and their sympathizers grow too impatient with the Saudis. In any case, the bilateral relations between the United States and Saudi Arabia will deteriorate and the rapidly increasing price of oil will have magnified the international financial problems of OECD and LDC countries to the point of seriously stunting economic growth. The main positive influences will be in the wealth of the oil producers and in the stimulation of the production of altenative sources of energy.

SCENARIO THREE: HIGH OECD ECONOMIC GROWTH WITH A RESOLUTE ENERGY POLICY

Demand for OPEC Oil

Economic growth picks up and an active energy policy is implemented by the OECD countries. The aims are to check dependence on oil imports through conservation and the expansion of domestic energy supplies and, at least implicitly, to retrieve part of the oil rent from OPEC. Rates of economic growth are 4 percent in North America and Western Europe and 6 percent in Japan. The energy policy reduces the energy coefficient to 0.7 in North America and 0.8 in Western Europe and Japan. The rate of growth of energy production is 4 percent in North America and 7 percent in Western Europe. In Japan energy consumption from domestic production and nonoil energy imports increases at a yearly rate of 9 percent. The rest remain the same as in the previous scenario, with a rate of economic growth of 5 percent, an energy coefficient of 1.2, and an aggregate growth rate of energy production of 6 percent.

The OPEC Response

The OPEC countries, fearing the adverse impact of the OECD energy policy on the demand for oil, reduce the growth of their imports after 1980 to only 5 percent a year in the countries of the first group, and 10 percent a year in the countries of the second group. Several of the countries of the first group produce at less than full capacity. As in the previous scenario domestic oil consumption increases at an annual rate of 8 percent and exports other than oil grow at an average rate of 10 percent a year.

119

TABLE 27

Scenario Three: Energy Consumption and Production in the 1980s and 1990s, in Millions of Metric Tons of Oil or the Equivalent

	1975	*1980*	*1985*	*1990*	*1995*	*2000*
Total Energy Consumption						
North America	1,894	2,278	2,615	3,002	3,447	3,957
Western Europe	1,138	1,374	1,608	1,883	2,204	2,580
Japan	329	435	580	696	879	1,112
OECD total*	3,361	4,087	4,774	5,580	6,530	7,648
Rest†	752	1,006	1,347	1,802	2,412	3,227
World‡	4,113	8,093	6,120	7,383	8,942	10,876
Domestic Output of Energy						
North America	1,599	1,945	2,367	2,880	3,504	4,263
Western Europe	521	731	1,025	1,437	2,016	2,828
Japan§	89	137	211	324	499	767
OECD total*	2,209	2,813	3,602	4,641	6,019	7,858
Rest†	499	668	894	1,196	1,600	2,142
World‡	2,708	3,481	4,496	5,837	7,619	9,999
Oil Imports						
North America	295	332	248	122	−57	−306
Western Europe	617	643	583	445	188	−248
Japan	240	298	340	371	381	344
OECD total*	1,152	1,274	1,171	939	511	−209
Rest†	253	339	453	606	811	1,086
World‡	1,405	1,612	1,624	1,545	1,323	876

*Excluding Australia and New Zealand.
†Including Australia and New Zealand.
‡Excluding the Soviet Union, Eastern Europe, and China.
§Including consumption of imported energy other than oil.
SOURCE: Author's projection.

120

TABLE 28
Scenario Three: OPEC Oil Exports in the 1980s and 1990s,
in Millions of Metric Tons

	1975	1980	1985	1990	1995	2000
The Distribution of						
Oil Exports						
Group I	651	857	808	681	516	294
*Group II**	754	756	816	865	807	583
Saudi Arabia	544	532	593	644	589	369
Saudi Excess Capacity						
at Different						
Production Ceilings						
750	206	218	157	106	161	381
1,000	456	468	407	356	411	631
1,250	706	718	657	606	661	881

*Includes Saudi Arabia.
SOURCE: Author's projection.

Price Politics

The low level of demand on the world oil market strengthens Saudi Arabia's position in OPEC and also creates serious tension in OPEC. The OPEC countries have little freedom of action as their trade balances become negative. There is great pressure on Saudi Arabia to accept a price hike. As the swing producer, Saudi Arabia has taken responsibility for balancing supply and demand to keep oil prices constant. As a result, however, Saudi Arabia might also have a negative balance of trade between 1980 and 1985. Consequently, the constant price of oil no longer serves Saudi Arabia's interests, and the position of OPEC has been further damaged by OPEC or IEA attempts to retrieve part of the oil rent. Thus Saudi Arabia agrees to a major price rise. The view prevalent in OPEC countries is that constant oil prices have hurt them and helped the OECD countries. They want to get as much as possible from oil while they still can, doubling the price in the early 1980s.

Political Tensions

Expanded energy production and conservation in the OECD countries is costly, but vulnerability to OPEC declines. Investment in the energy sector increases several times, new energy is expensive, and the conservation measures mean more investment and higher energy prices. There is a high import duty on oil from outside the OECD area. High taxes on energy consumption are designed to promote conservation and to finance some of the necessary investment. There are numerous direct restrictions on the uses of energy, such as the direct control of prices to prevent the high costs of energy from being offset through price increases by industry. Many of the schemes require subsidies that are followed up with strict governmental controls. The mobilization of capital for energy development limits investment in the rest of the economy and slows the growth of consumption. The energy effort is coordinated by the IEA, under United States leadership, and some financial assistance is channeled to energy projects in poorer member countries. United States coordination of the energy effort is also designed to foster political unity among the member countries.

In the OECD countries the principal effect of the energy effort is greater governmental intervention in economic life. Because of this and the high costs of energy self-sufficiency there is growing opposition to it in the OECD countries. Opposition comes from trade unions, which desire a higher growth of private consumption, and from the parts of industry that suffer from the capital squeeze. The wisdom of the effort is questioned because of the domestic sacrifices and the difference between the internal IEA energy price and the international oil price.

The LDCs are affected in two opposite ways. The relatively moderate oil price enables some LDCs to increase oil imports without inordinate strains on their current account balances. On the other hand, the strict energy programs in the OECD area have adverse effects on the exports of most LDCs, on the demand for raw materials, and on the volume of economic aid. On the whole, the LDCs are in a less desperate situation than with the high demand scenario, so the spread of nuclear technology is also more limited. The position of the poorest LDCs, however, is still precarious because of their inability to pay for energy and the loss of aid.

In the OPEC countries the principal effect is increasing political frustration. This scenario could potentially be very disruptive for the cartel. In desperation countries might try to underbid each other, and the cartel could collapse. On the other hand, OPEC might react collectively with a drastic price increase, hoping to break up IEA solidarity by offering special agreements to countries that abandoned the common import duty on non-IEA oil. Such a move would place the OECD governments in a difficult dilemma. The new international price of oil plus the import duty would probably create serious economic and political problems in the consuming countries. Abandoning the import duty would be a way out of these problems, but it would hamper the transition to other forms of energy and would dilute consumer solidarity.

THE ILLUSIONS

One of the most important negative effects of the Second Oil Regime is that it distorts our perception of the urgency of the world's energy problems. Since the oil crisis of 1973–1974 the world economy has adapted itself to higher oil prices and to the new structure of the oil market. Consequently, in the late 1970s there is a semblance of stability and a tendency among the public and the politicians to ignore the serious structural deficiencies of the oil market, both in its resource base and in its political organization.

There is a reluctance to accept the fact that the world, for the first time since the beginning of intensive industrialization, is facing increasing real costs for energy, and that the 1980s and perhaps the rest of this century will be a period of expensive and possibly scarce energy. There is still a widespread belief that there are no real supply constraints for energy and that those supply problems that exist are contrived by the oil industry or OPEC. As a result, there is a reluctance to conserve energy.

This is often linked to a belief that access to cheap and abundant energy is a natural right.[11] These opinions are much more popular in the United States than in Western Europe and Japan. This is understandable, given the past energy abundance of North America. But today, in a different historical situation, these opinions are irresponsible and potentially dangerous for North America and for the world. As a result, energy conservation and the expansion of domestic energy supplies are impeded, and oil imports are used to solve domestic energy problems.

The growth of United States oil imports has in the 1970s been the most dynamic factor in the world oil market, but this should not continue for long. First, the increasing consumption of foreign energy resources by a nation with a comparatively high level of energy consumption and large domestic energy resources is likely to cause political tensions with close allies and trading partners.[12] Second, by ignoring the basic energy problem, the United States is accentuating the descrepancy between the technological horizon and the market horizon for energy and consequently is contributing to the movement toward energy scarcity.

The oil-exporting countries have made no commitment to meet all of the world's demand for oil. They cannot be expected to indefinitely absorb the difference between energy demand and energy supplies for the rest of the world. If OPEC puts a ceiling on its supplies, major crises could develop among the consumers and between the OECD countries and the OPEC countries. A major factor in averting or in causing such shortages is the American demand for oil imports. This gives the United States a critical responsibility in the world oil market. Thus its energy policy is also a concern for the rest of the world.

In the United States there is a widespread belief that the Western world can muddle through the difficult energy situation in the years ahead, and that this is the most desirable solution because it avoids any formal commitments. However, it should

[11]William H. Miernyk, "Regional Economic Consequences of High Energy Prices in the United States," *The Journal of Energy and Development*, Spring 1976, p. 222.

[12]Seyom Brown, *New Forces in World Politics*, Brookings Institution, Washington, D.C., 1975, p. 36ff.

be clear from the three scenarios presented above that muddling through is neither without serious costs or without considerable risks.

A SUMMARY OF THE ISSUES AT STAKE

The 1980s and probably also the 1990s are likely to be a period of expensive energy. The real price of oil might perhaps double or even triple before the process of bringing forth alternative energy sources is established. If the present conditions prevail and demand for oil continues to grow, chances are that there will be one big price hike rather than a gradual rise. Under such circumstances, the new price will produce serious problems both for the consuming countries and for the world economy. This, in turn, will have a negative effect on the oil producers as well. In addition, oil and energy supplies will be scarce. Such developments clearly have serious short-term and medium-term implications for oil consumers, and equally important long-term implications for oil producers.

For the industrialized consumers oil supplies are not secure, because eventually incremental oil supplies from OPEC might not be readily available at any price. In addition, alternative energy supplies are not secure either, because their development is being retarded by circumstances on the oil market. The possibility of physical shortages in addition to the problem of paying for expensive oil limits the prospects of economic growth in the industrialized consuming countries. Also, the possibility of a dramatic price jump threatens economic stability because it could easily provoke a new, strong inflationary wave. The domestic political scene in most OECD countries could easily become unstable, with a high degree of polarization because of economic setbacks, inflation, and unemployment. Many countries might even opt for restrictive trade policies, reversing the liberalization of world trade. Finally, a tight world market would most likely create strong tensions between the United States and its major allies, with the United States, Western Europe, and Japan competing for favors from the oil producers. Thus a tight oil market could fragment the Western alliance system, economically and politically.

For the LDCs the most frightening probability is that economic development would be completely blocked by a new oil price jump and some of the poorest countries would experience serious setbacks. This could have important political effects. Moderate regimes might be replaced by new groups much more in opposition to Western interests. The LDCs could blame the new oil price hike on the OECD countries because of their excessive consumption, and it could be hailed as a new victory of the South against the North, even if it does hurt most LDCs.

For almost all oil exporters petroleum is the only significant national asset. They may achieve short-term gains from a new sharp increase in the price of oil, but they are also increasingly affected by the functioning of the world economy, and in particular by the economies of the OECD area. Therefore, they will suffer in the long run. Their continued economic growth depends on the long-term value of oil and their ability to diversify their sources of income. In addition, OPEC is more immediately threatened by an open conflict between its members over prices and supplies, which could disrupt or even destroy the cartel. The unity of the oil producers might also be disrupted by tensions or conflicts with the LDCs on questions of oil prices and the direction of North-South relations.

THE POTENTIAL FOR CHANGE

Given its political instability and its basic structural problems, the Second Oil Regime is not likely to last for a long time. Regardless of how fast prices rise, the Second Oil Regime will reach a crucial turning point when world demand for oil approaches the level of supplies that the OPEC countries are willing and able to provide.[13] As the scarcity of oil becomes increasingly apparent, the Western alliance system will be strained. West European and Japanese criticism of American energy behavior will be then

[13]Dankwart A. Rustow, "U.S.-Saudi Relations and the Oil Crises of the 1980s," *Foreign Affairs*, April 1977, p. 508.

openly voiced, identifying the United States as the world's number one energy problem and thus the country to blame for the crisis. This might be followed by the negotiation of important bilateral deals between OPEC members and West European countries and Japan, which would weaken the position of the United States in the OECD context.

In this situation the United States government, particularly if it has had few successes in implementing its energy policies at home, would attempt to act internationally. The Americans would have the choice of either trying to organize a front of consuming countries with an aggressive policy toward the oil producers or trying to lead the world toward some kind of global energy agreement.

The first option would mean an open conflict with OPEC designed to pressure the oil producers into providing the OECD area with more oil on better terms. This, however, would only postpone the world's energy problems because it does not begin to address the problem of a transition to alternative energy sources. The second option could forge a new oil regime stabilizing the world energy market. Such an agreement would be costly, but it would provide long-term security for producers and consumers. The outcome to a large extent depends upon the choice of policy by the United States. However, other OECD countries with important interests in common with OPEC, Great Britain and Norway, could also play an important role as mediators between the two groups of countries.

It is far from certain that the first option, an aggressive stance by the United States, would be successful. A military occupation of important oil fields is unrealistic and farfetched. It is militarily and logistically complex, and certain to foster tremendous political instability. A more feasible approach would be through IEA unity and the application of economic and diplomatic leverage. Given the differences within the OECD area, the organization, maintenance, and leadership of such a bloc are obviously difficult. If it is cohesive, its bargaining position in the world oil market is definitely weak relative to OPEC. The IEA's only hope lies in a tremendous energy self-sufficiency program that could give it some real bargaining power. This, of course, is not a short-term strategy, and there is no assurance that it will be successful. In the meantime the oil producers can raise prices and frustrate its success by making bilateral deals with individual OECD countries.

It might be argued that a global energy agreement is unnecessary, that it would be just another international bureaucratic structure, and that the market is likely to find its own solutions. Historically, most predictions of resource scarcity have proven to be false,[14] and today the world is better prepared than ever to respond to the depletion of important resources with technological innovation.[15] Changing technology is a solution typically worked out in a decentralized way and stimulated by the market. Consequently, there is a risk that an international agreement, by interfering with the market, could also impede technological change.

The market solution has obvious costs. Historical experience indicates that it takes from 30 to 50 years for a new source of energy to become dominant in the fuel mix of industrial societies.[16] During this period of adaptation, the world runs the risk of a prolonged economic slowdown. Massive conservation of energy may well be theoretically possible[17] but past experience indicates that energy conservation is difficult to bring about, is costly, and will only be effective after a long time. Moreover, the solutions worked out by the marketplace are unequal; they tend to favor the strong, the well organized, and the resourceful to the detriment of the weak, the less organized, and the poor. A major aspect of the world's energy problem is that the poorer countries, both the LDCs and the less well endowed OECD and OPEC countries, are hurt particularly badly. A solution worked out by the marketplace alone would certainly increase economic

[14]John M. Blair, *The Control of Oil*, Pantheon Books, New York, 1976, p. 15ff.

[15]Stewart S. Herman and James S. Cannon (eds.), *Energy Futures: Industry and the New Technologies*, Ballinger, Cambridge, Mass., 1977, p. 10ff.

[16]Hannes Porias, "Alternate Sources of Energy," in Ragaei El Mallakh and Carl McGuire (eds.), *U.S. and World Energy Resources*, ICEED, Boulder, Colo., 1977, p. 84.

[17]Joel Darmstadter, "Conserving Energy: Issues, Opportunities, Prospects," *The Journal of Energy and Development*, Autumn 1976, pp. 1–12.

inequalities between nations. The United States, Norway, and Saudi Arabia might benefit while most other countries would suffer. This could also have negative economic effects in the long run for the richer countries, and it would certainly create international political tensions.

The international energy market is already far from the free-market ideal of neoclassical economics. A strong cartel and government intervention give energy politics a decisive influence in shaping the market. Therefore, a political solution in the form of an international energy agreement would not be a radical step in relation to the well-established practices of the international energy market. An agreement could provide a more rational and coherent framework for the functioning of the market than is provided by the present cartel and the often uncoordinated interventions by governments.

At least three basic sets of conditions must be satisfied for an international oil agreement to work. First, as a basis for negotiation, the main participants in the world oil market, the industrial consumers, the producers, and the developing consumers must agree that the present organizational pattern of the oil market is unsatisfactory. Among the oil producers and the LDCs interest in an international energy agreement has been clearly expressed, but it has been linked to economic development and the management of raw-materials markets. OPEC countries are interested in an energy agreement only if it is achieved in the context of North-South relations. The group of developing countries at the Conference on International Economic Cooperation in 1976–1977, which included some OPEC countries, refused to enter into any kind of agreement guaranteeing energy supplies and prices unless there was a general agreement on economic development and raw materials. Saudi Arabia expressed its dissatisfaction in the wake of CIEC's failure by raising its oil prices by 5 percent.

Among the industrialized consumers, there is currently little willingness to accept Third World demands as a price for an energy agreement. There is also a diversity of opinion on the need for an international energy agreement, with Western Europe being more favorable than Japan or the United States. Therefore, either the OPEC countries must abandon the link between energy and LDC demands or the industrialized consumers must accept this link to make an agreement feasible.

Given the power relations within the world oil market, it is unlikely that the OPEC countries will abandon their position. The OPEC countries improve their bargaining position and get an added demographic base for their demands through their association with the rest of the Third World. If they were isolated politically from the LDCs they would be weaker in a world composed predominantly of oil-importing countries. As the bargaining position of the OPEC countries improves and that of the industrialized importers deteriorates, because of their increasing dependence on imported oil, there is little chance that the OPEC countries will soften their position. On the contrary, the OECD countries will probably have to back down in order to get an energy agreement. In any case, an energy agreement excluding the LDCs completely is both undesirable and politically hardly possible. Therefore, the understanding by the OECD countries that OPEC and the LDC links are strong, that their own bargaining position is gradually deteriorating, and that an international oil agreement is imperative will provide a basis for negotiation.

The second condition is that there must be an initial consensus on a few main principles:

- The OECD and the OPEC countries are dependent on each other for their economic health. Together they are responsible for the development of the world economy and the world's energy supplies.

- The depletion of the world's finite oil resources should be synchronized with the needs of consumers and the development of alternative sources of energy to provide a smooth transition from oil into substitutes.

- The oil-exporting countries should increase their output of oil up to an agreed-upon limit. In return, they should have greater opportunities for developing alternative sources of income, in the form of industrialization and foreign investment. In addition, the financial surpluses from international petroleum trade should be recycled according to mutually satisfactory guarantees.

- Cooperation between OECD and OPEC countries should be encouraged within a framework of North-South cooperation, for example, by promoting the transfer of technology and liberalizing access to OECD markets for oil exporters and LDCs.

The third condition is that the interests of the various parties involved be harmonized as much as possible as a basis for a successful agreement. This essentially means that possible conflicts related to specific oil interests should be offset by more general mutual concerns. Oil is a clear source of potential conflicts, but the more general long-term interests of the parties involved are basically convergent. All parties have a strong desire to avoid international turmoil, economic recession, a breakdown of the global economy, an economic and financial setback in the LDCs, an extreme radicalization of the oil exporters, or Western intervention in an oil-producing country. The fact that the more general political interests are more convergent than the particular oil interests, such as prices, supplies, and consumption patterns, allows for some flexibility in building a compromise between the two sides. This kind of cooperation is the key to a viable Third Oil Regime.

SEVEN

Patterns of Cooperation

The continuing dependence of the OECD countries on OPEC oil implies certain rational policy objectives for their relations with the OPEC countries:

- Confrontation and friction should be avoided in matters of immediate concern because in the short run the OPEC countries are now in a superior position given their financial surplus.
- The short-run asymmetry should be offset by developing common interests in matters of long-term concern.
- OPEC should be encouraged to develop a pricing policy that permits a more efficient allocation of the world's economic resources and gradually stimulates the development of alternative sources of energy.
- The OPEC countries should be encouraged to increase their production of oil and to invest in economic development and foreign assets rather than invest in oil in the ground.
- The OPEC countries should be encouraged to use the income from oil in a way that is beneficial to the world economy and to its prospects for growth.

The OECD consumers must recognize that their dependence on oil imports implies an increasing transfer of real resources from them to the OPEC countries. Only by accepting this fact can they influence the form and timing of the transfer. Essentially

133

the options are either a massive transfer over a relatively short period of time or a more gradual transfer over a longer period.

For the consumer countries a deferred transfer is preferable because it allows time to expand the productive base from which goods and services for a future transfer can be produced. There is widespread concern in the West over the OPEC surpluses but the preference in many quarters for absorbing them by an immediate growth of exports to OPEC countries is specious. A rapid transfer, in the form of a high level of exports to OPEC countries, means that resources will not be used for consumption and investment at home. Deferring the transfer through debts to OPEC countries gives cartel members a claim on future goods and services, and in the meantime resources can be used for OECD domestic economic expansion. Clearly, the form and timing of the transfer are of basic importance to the economic health of the consumer countries.

A major reason for the explicit preference in many OECD countries for a rapid transfer of resources is the lack of an international means to accommodate surpluses and deficits caused by the rising price of oil. Most countries seek to avoid large deficits, and there is no international framework for accommodating the inevitable deficits that result from extensive oil imports. In addition, most OECD countries do not receive substantial financial investments from the surplus exporters. To a large extent the OPEC surpluses have been placed in short-term deposits in the United States. This has created a considerable transfer of money to the United States from the poorer OECD consumers, particularly in Western Europe, improving American flexibility at the expense of other OECD countries. With an international framework for accommodating surpluses and debts related to oil it would be easier for the poorer OECD countries to accumulate deficits and maintain economic growth.

OPEC dependence on the OECD area as a market for its oil, as a source of imports, and as a place to invest its financial surpluses implies certain rational objectives for the policies toward the OECD countries:

- Confrontation and friction should be avoided in matters of long-term concern because in the long run the position of the OECD countries is improving.

- The long-run asymmetry should be offset by using the short-run superior position to develop common interests.

- The OECD countries should be encouraged to develop a combination of economic policy and energy policy that permits stable economic growth together with a gradual development of alternative sources of energy, giving a fair return to OPEC investment and assuring future markets for OPEC oil without undue pressure upon oil reserves.

- The OECD countries should be encouraged to accept investment and nonoil imports from OPEC countries.

- The OECD countries should be encouraged to develop monetary institutions and policies that will accommodate OPEC surpluses in a way that is beneficial to the world economy and its prospects of growth.

The OPEC countries must see that their economic development and the transfer of real resources that they can earn from their oil depend, to a certain extent, on the health of the OECD economies. This influences the form and timing of the transfer. Like the OECD countries, the OPEC countries have the option of either a massive transfer over a short period of time or a more gradual transfer over a longer period. The OPEC countries have an interest in a rapid transfer because it can be used to expand the domestic economy and thus increase, and possibly diversify, future production. The optimal use of oil revenues, however, is limited by domestic constraints in many OPEC countries, and not only in those with large reserves and small populations. After a point, with rapidly increasing levels of investment in domestic development, the rate of return decreases quickly.[1] Pumping too much money into the domestic economy can be harmful in the long run. It distorts the growth process at the microeconomic

[1] Anwar Jabarti, "The Oil Crisis," in Mallakh and McGuire (eds.), *U.S. and World Energy Resources*, p. 131.

level and thus reduces efficiency and creates bottlenecks, in addition to generating general inflationary pressure.[2]

A rational strategy for development requires planning, coordination, education, and balanced growth. A major task is to diversify the economy and improve the efficiency of individual firms. The supply of cash is only one ingredient; if it is used too liberally it can complicate this task instead of facilitating it. Related to this, at the government level, is the risk that easy access to capital can encourage involvement in badly planned industrial projects, with subsequent losses. The proposed large-scale involvement of Middle Eastern oil producers in refining and petrochemical industries, with at best uncertain prospects of return, provides an example.[3] Thus, for the OPEC countries, after a certain level has been reached, a deferred transfer of income may also be preferable.

Uncertainty over the future of the oil market and, in particular, uncertainty about the returns on foreign assets are probably to a large extent responsible for the desire in many OPEC countries for an immediate transfer of resources. This is due to the absence of an international framework for accommodating the surpluses and investment policies of oil producers. Without some stability, fears could develop in OPEC countries that the returns on their investments in the OECD area are not adequate. This could easily lead to more restrictive policies of oil production, to the detriment of the OECD economies and perhaps against the long-term interests of the OPEC countries.

Recent experience indicates that the return on OPEC assets held in OECD countries has not exceeded the general rate of inflation and that export prices are inflated relative to the general price level, thus giving a lower return to the OPEC countries than they would receive if they kept their oil in the ground and cut back the real transfer of resources. These gains can be seen as an advantage for the OECD consumers only from a narrow short-term perspective. The loss of potential earnings is bound

---

[2]Ibid., p. 130.
[3]Louis Turner and James Bedore, "Saudi and Iranian Petrochemicals and Oil Refining," *International Affairs*, October 1977, p. 575.

to be a growing concern to the OPEC countries, and if it is not mitigated it could lead to a decline in future oil supplies, which would harm the OECD economies.

As was shown in the scenarios, pressure on the price of oil can develop both from OECD-area demand and from a high import growth rate in the OPEC countries themselves. Consequently, both consumers and producers have an interest in a gradual transfer of resources over a long time. However, because of the lack of security on either side, they both tend to pursue divergent policies that compromise their long-term prospects for economic growth.[4] The dilemmas of the producers correspond closely to those of the consumers, and their policy objectives are quite complementary. This makes an international settlement a rational solution. It could give both sides the security they need to pursue policies that are in each other's best interest.

POLITICAL TRADE-OFFS

In order to be effective, any negotiated energy agreement must include formal mutual guarantees that secure the basic interests of both OECD and OPEC countries. This suggests that both parties make political sacrifices, reducing some of their freedom of action.

The OECD countries have a vital interest in increasing supplies of oil and avoiding the risk of a new price shock. This implies the accelerated depletion of OPEC oil reserves. To get a formal guarantee for this, they would have to offer the OPEC countries preferential treatment in matters of investment, trade, and transfer of technology. For example, regular government bonds can be eroded by inflation, but a new type of government oil bond might be created that provides OPEC investors with more security. In addition to these benefits the oil exporters should be allowed to participate financially in the energy sectors of the OECD countries, which would involve them in the development of alternative sources of energy. These kinds of concessions

[4]Jabarti, *The Oil Crisis*, p. 132.

mean that the OECD countries would have to accept a greater OPEC presence in their own economies, and in economic, financial, and trade policy take OPEC interests more explicitly into account. This might seem to be a tough bargain, but something along these lines is probably necessary to induce the oil exporters to accept an agreement that in essence implies relinquishing the oil weapon.

The OPEC countries have a strong interest in synchronizing the depletion of their oil reserves with their economic growth and, in particular, their development of new sources of income. The best use they could make of the oil weapon would be to become economically developed and to obtain long-term advantages for themselves and the Third World as a whole.

On both sides there are likely to be serious doubts and strong opposition to an oil agreement. For the OECD consumers it means institutionalizing their dependence on OPEC oil, and for the OPEC countries it implies giving up their newly acquired freedom of action in oil policy. Therefore, in order to be politically acceptable, an oil agreement must give long-term economic advantages to both sides that offset these political costs.

The formulation of an agreement that satisfies each side must take in a wide range of oil and energy problems. Consequently, I propose a package solution, a set of four different agreements, each one dealing with one specific set of problems, all of which are linked together. The following agreements are proposed:

- An oil price/supply agreement
- An energy/participation agreement
- A finance/investment agreement
- A trade/technology agreement

THE OIL PRICE/SUPPLY AGREEMENT

The main goals of the oil price/supply agreement are to bring the trends of energy demand and energy supply closer together by stimulating energy production and conservation, to gradually relieve the pressure on oil, and to organize a slow increase in the price of oil to prevent a new price shock. Ideally, the price of oil should increase as a function of the growth of demand pressure, and the price should reach the level of the cost of substitutes when pressure on available oil resources reaches a certain point. In practice this means reaching agreement on the critical price point discussed above and making the increases in the price of oil correspond to the growth of demand. Both sides would have to agree that up to a certain level oil would be made available by the producers. The price should be linked to a reference crude oil price, as is the case today in OPEC, with the market working out differences around the price that are due to quality and transportation costs. The aim then is to institutionalize a price rise over time that will stimulate conservation and the development of alternative sources of energy.

In spite of its apparent simplicity the principle of linking the price to the growing demand for a finite resource raises several pertinent questions: What is the realistic cost level of alternatives that should be used, given the far from perfect competition in the energy sector? How finite are the world's oil resources and at what point does the pressure on them justify pricing according to the cost of substitutes? How should the increasing oil rent, resulting from the gradual price rise, be distributed?

These questions are basic and there are no simple answers. The most equitable solution may be to let an internationally appointed team decide upon the cost level of alternatives that would serve as a reference point. It is much harder to agree upon the estimated oil reserves at present and projected prices and technology. Consequently, deciding when the critical point is reached, that is, determining the time when the pressure on reserves is too great, is extremely difficult. If pricing and reserve estimates are linked, this is certain to influence the behavior of governments and companies with oil. Most would probably try to underestimate their reserves, hoping for higher prices. A few, like Saudi Arabia, might be induced to do the contrary, hoping to delay the price rise. In any case, the search for oil would be stepped up but not the publication of results. Thus the attempt to establish the price and timing of the transition to other energy sources is not only difficult to calculate, it is also troublesome because it could distort the behavior of governments and firms.

One possible solution is to agree that the real price of internationally traded oil increase at a rate one and a half times the volume of oil imports to the OECD area. In return, the OPEC countries would collectively guarantee supplies up to some agreed-upon point. This would mean, for example, that when the imports increase by, say, 50 percent in relation to current net oil imports into the OECD, real prices for oil would be 100 percent higher. The rate of the price rise on oil each year would then depend on the growth of imports to the OECD area and would reflect a growing pressure upon the finite resource. If imports were stagnant or declined, the price would be kept constant. As a result, those governments and companies with oil would be encouraged to supply the market with as much oil as the OECD area demands because only in this way could they get the highest possible price. Hoarding the oil would cause prices to stagnate. This would make the task of determining the timing and level of the critical price a less volatile issue.

This oil price/supply arrangement would imply some form of indexation, as the real price of oil must move with the pressure of demand. Technically, indexation is very complicated.[5] Basically, it should reflect the inflation of prices on OPEC imports from the OECD area. Such an index could be made, but it would take a long time. A less than perfect solution is to use the International Monetary Fund (IMF) price index on exports of goods from 14 industrial countries.[6] This, however, excludes inflation on services imported by the OPEC countries, such as vital management skills. Some kind of an index would have to be agreed upon, and this technicality ought to receive close attention so that it is not an obstacle to the negotiation of the agreement. A system of indexation would also make OECD consumer governments quite wary of inflated export prices, because this inflation would raise the price of oil.

---

[5]Helmut A. Merklein, "Indexation of Oil Prices," in Mallakh and McGuire (eds.), *U.S. and World Energy Resources*, pp. 165–180.

[6]Ibid., p. 169ff.

The agreement only fixes the international trading price for a reference crude, which gradually increases the incomes of the oil exporters but leaves the consuming countries to tax uses of oil as they see fit. An essential element of the agreement is that in return for a gradually rising price the oil exporters collectively guarantee supplies. This could create serious tensions among the oil exporters over the distribution of output and income. In practice, this could work as now, with Saudi Arabia assuming the role of residual producer. Given the rising price, oil exporters with small reserves might want to reduce output, in order to enjoy higher future revenues from their limited oil reserves. But the price rise is going to be gradual, and given their large income requirements, most would probably produce up to capacity. Also, as price increases under the agreement are linked to an expansion of international oil trade, withholding supplies would have a negative effect on prices and incomes. In fact, the other OPEC countries would have an interest in Saudi Arabia expanding its supplies. A more explicit recognition of Saudi Arabia's residual role could, however, help stabilize the world oil market.

In the context of North-South relations the agreement should include compensation to the LDCs for the rising price of oil. Such compensation would be based upon the increases in the price of oil, and the LDCs would receive some proportion of this increase. The refund could take the form of credits that are based on an internationally accepted GNP index. Thus, with increasing GNP, the refund credit decreases. The credit would be granted as a long-term loan at low interest rates, so that in practice the LDCs would not pay their full oil import bills. In order not to place an excessive burden on individual oil exporters, the credit should be administered by an international financial institution, preferably IMF. This compensation to the LDCs would be an important relief for their balance of payments but would not be significant globally, as the oil imports of the LDCs are relatively small. The countries benefiting from this compensation would of course have to allow international inspections to ensure that they are not reexporting oil at a profit.

If compensation to the LDCs is not agreed upon, OPEC might

141

consider putting a development tax on its oil exports to industrialized countries.[7] The proceeds could be channeled to the LDCs through the OPEC Development Fund or existing institutions. A modest tax of this kind would generate substantial funds for development aid. For example, a tax of $0.25 per barrel would at present levels of oil trade generate $2.5 billion a year. Such a solution would certainly strengthen the alliance between OPEC countries and LDCs. Thus, to avoid such a tax, it is in the interest of OECD consumers to accept LDC price compensation.

If the oil price/supply agreement is not effective in bringing trends of energy demand and energy supply closer together— that is, if long-term price elasticities remain very low and a supply squeeze develops—a pricing agreement could be added. At present, this does not seem necessary, desirable, or feasible. The aim of the pricing agreement would be to discourage the use of oil where other forms of primary energy can be relatively easily substituted and to further stimulate the development of alternative energy sources. This implies giving different priorities to different uses of oil. For example, take the following ranking:

1. Highest priority and lowest price are for the use of oil for petrochemical purposes.

2. The use of oil for combustion engines should have a relatively medium price.

3. Lowest priority and highest price are for the use of oil for electricity and heating.

In addition, this optional pricing agreement could be extended to other forms of energy production. For example, the production of alternative sources of energy, whose development is seen as particularly desirable, could be encouraged through higher prices. The output of synthetic oil, from coal, oil shale, or tar sands, and oil and coal produced under adverse conditions would be promoted. The differentiation of oil prices by end uses could provide outlets for this higher-priced energy. In practice, this

[7]Ibid., pp. 179–180.

would require extensive controls and planning procedures and would only be feasible in a critical situation.

THE ENERGY/PARTICIPATION AGREEMENT

The main goals of the energy/participation agreement are to organize a common institutional framework between oil producers and consumers, to pool resources and channel efforts in the energy sector, and to give the oil producers real assets and a stake in the energy consumption of the consuming countries, especially in alternative sources of energy.

The best allocation of resources in the energy sector can be ensured by cooperation and, to a certain extent, through reintegrating the international oil companies. They are vital not only because of their technological, organizational, and financial resources; they are also valuable intermediaries in the international economy.[8]

The breakup of the integrated structure of the oil market in the oil revolution created serious problems for both sides. The international oil companies and several national oil companies of the consuming countries lost many of their sources of oil. Even though several companies made large profits on inventories during the oil crisis, many are now in an extremely tight financial situation. This is especially so for the Western European national oil companies.[9]

In recent years oil companies have increasingly moved into other sources of energy, such as coal, oil shale, and uranium. This type of diversification, called horizontal integration, has been defended because it centralizes capital and expertise, thus furthering the development of new sources of energy. It has been severely criticized, especially in the United States, because it impedes competition in the energy sector and could lead to substantial price increases for consumers. Because of rising costs

[8]Raymond Vernon, *Storm over the Multinationals*, Harvard University Press, Cambridge, Mass., 1977, p. 175ff.

[9]Jean Carrié, "Les Cinq Compagnies d' Europe occidentale et la crise énergétique," *Politique Etrangère*, no. 2, 1977, pp. 155–165.

and the financially strapped position of several companies, the process of horizontal integration seems to have slowed down. If unchecked, it would probably accelerate the development of new sources of energy, but at the same time it would create a strong international energy cartel and possibly price jumps. Restricting horizontal integration would reduce the risk of an international energy cartel, but at the cost of slowing down the development of new sources of energy, and thus also creating pressure for price increases. Thus the resources of the oil companies are needed for expanding energy production, but there is also a good case for stronger political control over them. For the oil-exporting countries, the breakup of the integrated structure has meant the loss of secure outlets for their oil and new problems of adapting to the market, as is reflected in plans of the oil-exporting countries for more extensive downstream engagement. In fact, the breakup created a new intermediate market in the oil industry, which produces additional uncertainties and potential economic losses to both sides.

The solution envisaged here is to use the international oil companies as the cornerstone of a new oil regime. Their expansion and involvement in other types of energy should be encouraged under closer public control. This can be organized through joint ventures and state participation by both consumer and producer countries in the international oil industry. I propose two alternative solutions, one moderate and the other more radical.

The moderate solution is to organize a system of service and participation contracts. Service contracts require the international oil companies to provide certain skills to the producer countries with payment in crude oil.[10] The participation contracts allow the national oil companies of producing countries, which lack downstream networks, to refine and market some of their crude oil through the organizational structure of the OECD-based oil companies. Compensation for this can be paid in crude, and, in recognition of this cooperation, the producing company is

[10]This solution would ensure more secure access to crude oil. With increasing prospects of real shortages, the physical flow of oil is likely to get a higher priority than the question of the flow of money.

144

represented on the board of the OECD-based company. This can be supplemented by the producing company using its crude oil to buy shares in firms with extensive downstream operations. Also, the various oil companies, both producing and marketing types, could work together in joint projects in consuming countries. Finally, a system of energy-planning agreements can be instituted between governments of the consuming countries and the private oil industry.

The more radical solution is to increase the capital of the international oil companies through the direct participation of consuming and producing governments or, indirectly, through their national oil companies. The expansion of capital would increase the economic resources available to the international oil companies, allowing for more horizontal integration. The participation by OECD governments directly or through national oil companies would give them more control over the oil business and the energy sector. The process of horizontal integration in energy and the increased economic strength of the international oil companies would be politically more acceptable if governments had this increased control. Participation by OPEC governments and their national oil companies would broaden their knowledge and involvement in downstream operations such as refining and marketing, give them secure outlets for their oil, and perhaps delay their projected building of refining and petrochemical industries. Last but not least, it would give OPEC governments a stake in alternative sources of energy and an even greater interest in the energy consumption of the OECD countries.

For the international oil companies the participation of OECD governments could mean closer cooperation with public authorities, particularly in the public planning process and in the development of other forms of energy. State participation by OPEC governments could make sources of crude oil more secure and would also mean closer involvement in OPEC planning. With several governments involved in each company, no single government could have exclusive control over a company. Thus the international oil companies would serve as both links and buffers between producing and consuming countries, their technical and human resources would find a wider field of application, and their economic resources would be considerably expanded. However, their operations would come under closer public scrutiny, and they would have to be more responsive to the concerns of governments and parliaments. This would be a minor disadvantage compared with the security of supplies and markets together with the new economic resources and functions.

THE FINANCE/INVESTMENT AGREEMENT

The main goals of the finance/investment agreement are to organize a framework that can accommodate surpluses and debts relating to oil, to offer OPEC countries attractive investment opportunities, and to stimulate oil production.

The basic principle of the agreement is that the deferred transfer of resources should be facilitated and that new investment opportunities in the OECD area should encourage OPEC countries to accelerate the depletion of their oil reserves within reasonable limits. In practice this means that investing oil revenues in foreign assets must be more attractive than keeping oil in the ground. The idea is to encourage OPEC members with large reserves and small populations to expand their production of oil and at the same time allow OECD countries to defer the financial impact of paying for this oil. This could obviously improve the position of the OECD members whose international financial position and prospects for growth are endangered by the rising price of oil.

The solution proposed here has three different aspects: (1) to improve facilities for short-term credits of oil importers, (2) to institutionalize a system for handling long-term debts and surpluses between oil importers and oil exporters, and (3) to promote direct investment by oil exporters in consumer countries.

Even a gradually increasing price for oil, as envisioned in the oil price/supply agreement, could create serious economic problems for OECD countries such as Denmark, France, Italy, and Portugal. It would place an additional burden on their balances of payments and also would inhibit economic growth. Granted, a gradual price rise is likely to be manageable for the OECD area

146

as a whole, but this is not the case for some individual countries. A gradual price rise might have only a limited real impact, but still it could be psychologically and politically damaging, thus threatening the economic growth and political stability of certain countries.

Needy OECD countries should have access to international credits to offset part of their rising oil import bill. These credits should not exceed a certain percentage of the oil imports bill, for example, 20 percent; they should be at a relatively low interest rate, for example, 6 or 7 percent; and they should be granted through an existing institution, preferably the IMF. Channeling the credits through the IMF is good because it establishes a certain neutrality. These oil credits should be refundable after about five years. If oil credits are not reimbursed after the five-year period, they should be converted into oil bonds, which are described below. The credits should be financed through contributions by surplus oil exporters and surplus oil importers. For example, given their surpluses, West Germany and Japan would be expected to contribute.

Gradually increasing oil prices and rising levels of oil trade are likely to magnify the more structural aspects of balance of payments deficits and surpluses. These problems underline the need for long-term credit solutions. In order to help defer OECD payments and stimulate oil production, the oil-consuming countries should offer special bonds to the oil producers.[11] These oil bonds should be issued under the auspices of an international organization, ideally the IMF, and they should be indexed as a guarantee against inflation and exchange-rate instability. Indexed bonds do not exist in the United States, but they are available in Europe.[12] The introduction of indexed bonds by the United States is a small sacrifice to make for stimulating the oil producers to increase output. The bonds ought to carry the current rate of interest to be competitive.

Within reasonable limits set by an international institution, a consumer country could issue oil bonds directly to its oil sup-

[11]Helmut A. Merklein and W. Carey Hardy, *Energy Economics*, Gulf Publishing Co., Houston, 1977, p. 106.
[12]Ibid.

pliers. If an oil-producing country prefers cash, it could sell the oil bonds back to the supervising international institution, which would then sell them to other countries that want to hold bonds. Also, as mentioned before, short-term oil credits could, within limits, be converted into oil bonds. Such a system of oil bonds would be an important contribution to international monetary stability[13] and would be an incentive to oil production in countries that might have otherwise opted to leave their oil in the ground.

Bonds alone hardly provide adequate compensation for the accelerated depletion of oil reserves.[14] In addition to bonds, the oil consumers must make real assets available to the oil producers. Direct investment by oil producers in the OECD area should be encouraged. Under the auspices of an international agreement the buying of real assets by oil exporters could be made more attractive. This can be done by removing limitations on the repatriation of profits and interest, by liberalizing investment regulations, by abandoning special taxes and restrictions on interest rates and returns paid to foreigners, and by ensuring that there are no official restrictions on normal international financial transactions.[15]

THE TECHNOLOGY/TRADE AGREEMENT

The main goal of the technology/trade agreement is to organize a framework for the economic development of oil-exporting countries, emphasizing diversification, modernization, and industrialization. This would tie imports growth to a process of structural change, eventually making the interdependence between OECD and OPEC countries more balanced.

The basic idea is that the transfer of technology from OECD to OPEC countries should be facilitated and that OECD-area

[13]Ragaei El Mallakh, Mihssen Kadhim, and Barry Poulson, *Capital Investment in the Middle East*, Praeger Publishers, New York, 1977, p. 163.

[14]Merklein and Hardy, *Energy Economics*, p. 106.

[15]Abdulaziz M. Al-Dukheil and Darwin Wassink, "Oil Production, Pricing and the International Financial System," in Mallakh and McGuire (ed.), *U.S. and World Energy Resources*, pp. 157–164.

markets should be opened up to a broad range of goods from the OPEC countries. In practice this means that OECD countries guarantee OPEC countries access to all modern technology (with the exception of nuclear military technology) at nondiscriminatory prices and that assistance will be provided for the transfer and initial operation of this technology. It also means a long-term trade agreement that gradually reduces OECD trade barriers to the industrial and agricultural exports of the OPEC countries.

The solution envisioned here has four aspects: (1) a system of joint projects for the transfer of technology at the government level, (2) the supervision of technology transfers at the corporate level, (3) a trade agreement providing markets for OPEC exports, and (4) the encouragement of regional economic cooperation.

Foreign assets can provide OPEC countries with a good source of income, but they are no substitute for domestic economic development. The majority of OPEC countries must industrialize immediately because of their limited oil reserves and large populations. Even the OPEC countries with the largest reserves have an interest in diversifying their economies, particularly through the development of energy- and capital-intensive industries.[16] The economic diversification and industrialization of the OPEC countries is important to the OECD countries too because it will eventually make the interdependence between the two groups more balanced.[17]

The development process in the OPEC countries will lead to higher standards of living, greater dependence on imports, and an increasingly differentiated relationship with the OECD countries. This will make it harder for OPEC countries to deny the closer community of interests between the OPEC and OECD economies. However, the rapid growth of OPEC imports from the OECD area does not, in itself, create balanced interdependence. The value of the imports is diminished by inflation in the OECD area and the low rate of return on new investments in the OPEC countries. A more balanced interdependence, benefiting

[16]Ibid., p. 159.
[17]U.S. Congress, Senate Committee on Interior and Insular Affairs, *Geopolitics of Energy*, 95th Cong., 1st Sess., 1977, p. 131.

both sides, can only be achieved through a structural change in the OPEC economies, which takes time and requires planning and coordination. Consequently, economic and social modernization are not a direct function of the growth of imports. In fact, a sudden upsurge in import growth could, in the worst case, provoke a backlash against development which would hurt the position of OECD countries. Therefore, it is in the interest of both sides to tie the oil revenue and the growth of OPEC imports to a more durable structural change.

The system of joint projects for the transfer of technology is basic to the structural transformation of OPEC economies. Through this system the government of an OPEC country could request from an OECD government a specific piece of industrial technology available in that country. It would then be the responsibility of that particular OECD government to arrange that the technology in question be made available on equal and non-discriminatory terms, and that during an initial period technical assistance would be provided. Of course, an exception would be made for nuclear military technology. In many ways this process is already taking place between OPEC and OECD countries, but without an international agreement terms vary considerably and equal, nondiscriminatory treatment is not always assured. An international agreement would involve governments directly and also give them more control over the terms of technology transfers.

The supervision of technology transfers would complement the OECD government guarantees by institutionalizing the process at the corporate level. In order to ensure equal and non-discriminatory treatment from OECD firms, their governments would be made responsible for what happens between companies. OECD governments would ensure that joint ventures between two or more companies from OECD and OPEC countries for implementing the transfer of technology would follow the same guidelines advocated at the national level. Thus companies from OPEC countries would be treated fairly by companies from the OECD area, in a way that encourages structural change in the economies of the OPEC countries.

A trade agreement should provide markets for the new industrial production of the OPEC countries, giving the oil producers new sources of income. The trade agreement would set a timetable for the gradual reduction of OECD-area import duties and other trade barriers against industrial products from the OPEC countries. Some agricultural exports could also be included in the trade agreement. An exclusive trade agreement between OECD and OPEC countries would certainly be seen as discriminatory by the non-OPEC LDCs. It is hard to say whether or not the OPEC countries would accept an exclusive bilateral trade deal with the OECD countries, because it could weaken their ties with the LDCs. This underlines the importance of the general North-South context. Here, for reasons to be spelled out below, it will be assumed that such a bilateral OECD-OPEC trade deal is politically feasible.

To a considerable extent, the ability of OPEC countries to absorb their financial surpluses is linked to their aid to non-OPEC LDCs and regional cooperation among OPEC and non-OPEC countries.[18] The OECD countries have a definite interest in developing the long-term basis for the absorption of surplus funds, and regional cooperation can be an important tool for this. It can also provide OPEC countries with a more stable and friendly economic and political environment. The West should therefore promote the development of the neighbors of important OPEC countries, particularly when this is partly financed by the latter.

The way to encourage these regional systems is to let non-OPEC countries that are taking part in regional economic cooperation with OPEC members enjoy the benefits of the technology/trade agreement. This means, in the case of the Middle East and North Africa, that Syria, Lebanon, Egypt, and possibly Tunisia and Morocco, to mention a few, should also be granted technology transfer and trade liberalization guarantees on an equal basis with the OPEC countries. The long-term impact

[18]Mallakh, Kadhim, and Poulson, *Capital Investment in the Middle East*, p. 49ff.

would be a more balanced economic development of the entire Middle East. In the long run this will not only reduce fears about the accumulation of financial surpluses from oil, but will also foster a more stable political environment and thus more secure oil supplies.

CHOICES

The four proposed agreements presented above as a package are cumulative. Indeed, an agreement could consist of the oil price/supply agreement alone, or it could include energy participation, finances and investment, or even the transfer of technology and trade. The more extensive an agreement is, the more difficult it will be to negotiate. But a package agreement like the one presented above could more systematically reconcile differences among oil producers and consumers.

The choices of scope and participation are of decisive significance. An agreement of limited scope, dealing with oil prices and supplies alone, would create serious problems for the poorer OECD countries, not to mention the LDCs. Its effectiveness in guaranteeing increasing supplies of oil is questionable because the OPEC countries would not receive benefits other than gradually increasing incomes, which could also be easily obtained without an agreement through a unilateral price hike. It is tempting for the OECD countries to push for a bilateral OECD-OPEC deal, but there is no reason to expect OPEC to accept such an agreement. By taking wider interests into account, a package solution could tie all parties closer together and thus provide more effective guarantees and greater stability. Furthermore, including the LDCs is advantageous because it builds a mediating third group into the system.[19]

There are also good reasons for keeping the agreement open to the Soviet Union, Eastern Europe, and China. The countries

[19]Edith Penrose, "Aspects of Consumer/Producer Relationships in the Oil Industry," in Mallakh and McGuire (eds.), *U.S. and World Energy Resources*, p. 27ff.

of Eastern Europe will increasingly need to import oil, which puts them in the same position as the Western Europeans. Through their participation in an oil agreement, their basic oil interests would be secured, and they would be brought more closely into the international community. East European participation in a package agreement—including energy participation, finance and investment, and technology and trade—would certainly be an economic and political innovation. If the Soviet Union becomes a net importer of oil in the 1980s, it would be imperative to have an international framework for solving issues related to oil, in order to avoid superpower competition for limited oil reserves. This would benefit not only the consuming countries but certainly the oil producers as well. The participation of China, as an oil exporter, could bring certain advantages to that country and stimulate the development of its oil exports.

INSTITUTIONS

An extensive international energy agreement such as the package deal presented here does not require a large international bureaucracy to be effective. However, some institutional innovation would have a positive effect on the behavior of the parties involved, contributing to the stability and longevity of the agreement. An oil agreement without its own organizational structure might deteriorate more easily than one with an institutional base and vested administrative interests working permanently to resolve conflicts. Therefore, the solution preferred here includes some new international organizations.

One option would be to create a joint agency under the auspices of OECD and OPEC to administer the agreements. The four proposed agreements need not be built into the charter of the institution. The new organization could be designed to act as a supervisor and administrator of all types of bilateral or multilateral energy agreements. The IMF could carry out the financial transactions involved in any agreement, but this would have a significant impact on the resources and functioning of IMF. Consequently, the leading surplus oil exporters, primarily Saudi

Arabia and perhaps Kuwait, should be granted permanent positions in the decision-making bodies of the IMF.

To have some effectiveness, UNEO should be able to influence national oil and energy policies. The minimum would be a regular examination and comparison of national oil and energy policies. This could be similar to the present OECD yearly studies and recommendations on economic policies of member countries. A stronger step would be for UNEO to link recommendations to material incentives and disincentives in the form of credits. UNEO might be granted wider powers in a tighter energy market because of the increased potential for conflicts over oil and energy. These could include fixing an international price structure for oil and other forms of energy according to end uses and the need to develop new energy sources. In the most extreme case the organization might even allocate scarce resources. Such immense powers seem neither desirable nor necessary in the present situation. However, now and in the 1980s, UNEO could play an important role in pooling resources and channeling efforts in the development of alternative sources of energy. It could also implement agreements on price compensation and energy assistance for the LDCs.

Given the global nature of the energy problem, perhaps a better method is to attach the agreement to the United Nations. This could be done through the creation of a new United Nations suborganization for energy, the UN Energy Organization (UNEO). Its aim would be to promote international cooperation on energy questions, regionally and between producers and consumers, and to encourage the formulation of an international energy resources policy. This policy would encompass the exploitation, conservation, and consumption of the world's energy resources and would attempt to ensure that energy resources are used rationally in the long run.

The organizational structure of UNEO might be built around a General Conference of member countries as its ultimate authority. This raises questions about the nature of membership. The main purpose of the package agreement is to satisfy the interests of all sides—oil exporters, industrialized consumers, and LDCs—but membership by blocs of countries is contrary to the principles of the United Nations. This could be overcome through an Executive Council, functioning much like the Security Council, whose membership would be equally divided between oil-exporting countries, industrialized consumers, and LDCs. This might pose problems in the definition of LDCs and industrialized consumers. A simpler solution would be to let membership on the Executive Council be equally divided between net energy exporters and net energy importers. This would create cleavages and alignments along functional lines different from the present alliance lines.

The Executive Council would appoint a Secretary General, heading a permanent Secretariat. The main tasks of the Secretariat would be to supervise the different agreements, compile statistics, and regularly inspect and compare energy policies. Attached to the Secretariat would be an International Energy Institute. Its main tasks would be to help countries with energy conservation and the development of new energy sources, especially those countries that lack technical skills.

COMMITMENTS

The package solution presented here would imply—even without UNEO—important commitments from each group of countries. The governments of the industrialized consuming countries would have to accept gradually increasing prices for oil and permit involvement by OPEC countries in downstream oil operations and in other forms of energy production in the OECD area. Financially, OECD governments would be committed to issuing special indexed oil bonds, and, industrially, they would have to accept investment by oil exporters on equal and nondiscriminatory terms. Furthermore, OECD governments would guarantee and supervise the transfer of technology and open their markets to OPEC goods other than oil.

LDC governments would also have to accept a gradually rising price for oil. Price compensation for these high oil prices would decline with higher levels of development, and those receiving compensation would have to agree to international inspection of their oil trade. These countries would also have to accept OPEC involvement in downstream oil operations and other energy projects.

The governments of the OPEC countries would be committed to supplying increasing quantities of oil, at gradually rising prices, up to an agreed-upon level. Financially, they would, to a certain extent, have to accept oil bonds as payment for their oil. Thus all three groups of participants must accept significant obligations, and this is certainly an obstacle to concluding an agreement. Nevertheless, the rewards for all sides could be tremendous.

THE CONDITIONS FOR A SETTLEMENT

The main argument in favor of an international energy agreement is that the energy problem is global. No country can solve it alone, and ultimately the dilemmas of the oil producers and consumers are complementary. The world oil market is characterized by a trend toward polarization between OPEC and IEA. This obviously increases the risk of a future confrontation, which would hurt all parties. However, this bilateral structure also carries the seeds of cooperation.

The need for cooperation is frequently voiced on both sides,[20] but there is no consensus on how to bring it about.[21] The main obstacle is probably the unwillingness of OECD countries to recognize the link between energy problems and the wider North-South context. This is the reason for the deadlock and failure of the CIEC in the spring of 1977. There are also other obstacles in the OECD countries. First, there is an unwillingness to realistically assess OECD-area dependence on foreign oil, which is compounded by the belief that the cartel will soon break down. Second, there are vested interests opposed to an energy agreement, in part because it would imply more public intervention in the energy sector. There are obstacles on the OPEC side as well. In the OPEC countries there is sometimes a reluctance to recognize how dependent the OPEC countries are on the eco-

[20]See, for example, U.S. Library of Congress, *Project Interdependence*, Washington, D.C., 1977, p. 75.
[21]U.S. Congress, *Geopolitics of Energy*, p. 23.

nomic health of the OECD area. OPEC perceptions are sometimes clouded by the dream of retreating to a pastoral past if things go badly.[22] Some members of the cartel make unrealistic assumptions about the strength and potential of the oil weapon, ignoring the fact that most of its strength resides in its not being used.

Thus there are extremists in both groups. However, as the oil crisis recedes into the past, there is a sobering on both sides and cooperation grows more popular.[23] It now seems possible to convince the vast majority of OECD governments that energy cooperation with OPEC countries is desirable. It is much harder to arrive at a consensus among them on prices or other specific elements of an agreement. There is a growing awareness among OPEC countries that their long-term development is closely linked to the fate of the Western economy and that cooperation could thus help both sides. It is not as well understood that cooperation implies giving up or limiting considerably the oil weapon. The task, therefore, on the OECD side is to convince the skeptics that an energy agreement is so important that it is worth paying higher prices for oil and making some concessions in North-South relations. The OPEC side must convince the skeptics that it is worthwhile to relinquish the oil weapon in return for substantial economic advantages and progress in North-South relations.

The current situation has definite risks for both sides, which underline the need for negotiations. For the OECD countries economic stability increasingly depends on oil supplies, and these in turn depend on Saudi Arabian policy choices. The decisions the Saudis must make become more difficult as pressure from both sides grows.[24] Little is known about Saudi Arabian politics. The country appears to be stable, but possible political changes are hard to predict.[25] It is clear that Saudi Arabia is still in the

[22]Ibid., p. 134.

[23]Ulf Lantzke, "International Cooperation on Energy—Problems and Prospects," *The World Today*, March 1977, pp. 84–94.

[24]Louis Turner, "Oil and the North-South Dialogue," *The World Today*, February 1977, p. 58.

[25]U.S. Congress, *Geopolitics of Energy*, p. 140.

process of adapting to its new situation, and there is no guarantee that there will be continuity in its oil policy. For the OPEC countries there is the danger that an economic crisis in the West might lead to more aggressive regimes that will take a tougher stance with OPEC. Furthermore, the demise of CIEC breeds pessimism on both sides and has stalled any overt progress toward an agreement.

The United States has a particular responsibility, as the world's largest consumer and importer of oil and as a leader of the OECD area. Rising oil imports are an increasing burden upon its balance of trade, and they thus contribute to the instability of the dollar on the foreign exchange market. This situation, combined with possible setbacks in domestic energy policy, could stimulate an American reassessment of international energy policy. One option for the United States is to offer the South an agreement on economic development on the condition that an oil agreement is reached. Another possibility is to present a whole package, similar to the agreements proposed here, plus a North-South deal. On the other hand, OPEC countries that are worried about their economic future and their dependence on the West might well take the initiative, on the condition that progress is made in North-South relations.

Once an agreement is established, its effective operation would essentially depend on a few key countries respecting it. Both Saudi Arabia and the United States must be strongly committed to the agreement for it to work. However, the agreement would not directly maintain the United States–Saudi bilateral relationship but rather transform it by giving Saudi Arabia a key role in handling both oil and finances. This is the crux of the proposed agreement. By institutionalizing Saudi Arabia's position, the agreement would create a framework for accommodating the pressures on the Saudis, and this would stabilize the oil market.

CONCLUSION

This look at the 1980s indicates that the future of the international oil market is riddled with uncertainties. There are many possible scenarios, but it is certain that the days of inexpensive oil are behind us. The economic boom from 1950 to 1973 was without precedent in the Western world, and it was literally fueled by rapidly increasing inputs of energy and growing Western dependence on imported oil. Given the future instability of the world oil market, especially if market forces and political competition are the principal means of resolving differences, economic growth is not likely to flourish as it did in the postwar period. A search for patterns of cooperation, as exemplified by the proposed agreement, is needed.

The international oil and energy agreement described above is unique. Traditionally, commodity agreements have broken down because of drastic changes in their premises. The package solution envisioned here entails some important self-regulating mechanisms. Through involvement in downstream oil operations and in the development of alternative energy sources, the OPEC countries would get a direct interest in the oil and energy consumption of the OECD area. OPEC members would thus gain more influence over trade-offs between energy consumption, pressure on their own oil, and the development of alternatives. In essence, the oil producers acquire more control over the transformation of the world energy market, which gives them a much greater interest in the stability of the system. As a result, the OECD countries are assured much more economic security. This intertwining of interests provides the basis for a new oil regime that is self-sustaining.

The main point of this prescription is that a negotiated oil agreement is a desirable and viable solution to the oil problems of both the OECD and the OPEC countries. The intertwining of interests in a package deal creates a new oil regime that takes into account the needs of all sides and thus helps provide a basis for stable economic growth. Granted, the political costs are rather high for all parties involved, but it is the claim of this study that these costs could be easily offset by the economic benefits for all participants in an international agreement. The feasibility of such an international agreement merits further exploration.

Selected Bibliography

Adelman, M. A.: *The World Petroleum Market*, The Johns Hopkins University Press, Baltimore, Md., 1972.

Blair, John M.: *The Control of Oil,* Pantheon Books, New York, 1976.

Brown, Seyom: *New Forces in World Politics*, Brookings Institution, Washington, D.C., 1975.

Carrié, Jean: "Les incidences de la crise énérgetique sur l'économie de L'Europe et des États Unis," *Politique Étrangère*, no. 1, 1975, pp. 85–97.

Chesshire, John, and Keith Pavitt: *Social and Technological Alternatives for the Future—Energy*, Science Policy Research Unit, University of Sussex, Brighton, England, 1977.

Chevalier, Jean-Marie: *Le Nouvel Enjeu Petrolier*, Calmann-Levy, Paris, 1975.

Darmstadter, Joel, et al.: *Energy in the World Economy*, The Johns Hopkins University Press, Baltimore, Md., 1971.

———, Joy Dunkerley, and Jack Alterman: *How Industrial Societies Use Energy: A Comparative Analysis*, Resources for the Future, Washington, D.C., 1977.

Doran, Charles F.: *Oil, Myth and Politics*, The Free Press, New York, 1977.

Eckbo, Paul Leo: *The Future of World Oil*, Ballinger, Cambridge, Mass., 1976.

Engler, Robert: *The Brotherhood of Oil*, University of Chicago Press, Chicago, 1977.

Fried, Edward R.: "World Market Trends and Bargaining Leverage," in Joseph A. Yager and Eleanor B. Steinberg (eds.), *Energy and U.S. Foreign Policy*, Ballinger, Cambridge, Mass., 1975.

Georgescu-Rogen, Nicholas: "Energy and Economic Myths," *Southern Economic Journal*, vol. 41, no. 3, January 1975, pp. 347–381.

Houthakker, Hendrik A.: *The Price of World Oil*, The American Enterprise Institute of Public Policy, Washington, D.C., 1975.

Jacoby, Neil H.: *Multinational Oil*, Macmillan, New York, 1974.

———: "Oil and the Future: Economic Consequences of the Oil Revolution," *The Journal of Energy and Economic Development*, Autumn 1975.

Jaidah, Ali M.: "Pricing of Oil: Role of Controlling Power," *OPEC Review*, June 1977.

Long, David E.: *Saudi Arabia*, Center for Strategic and International Studies, Georgetown University, Washington, D.C. 1976.

Mallakh, Ragaei El: *Implications of Regional Development in the Middle East*, Praeger Publishers, New York, 1977.

———, and Carl McGuire (eds.): *U.S. and World Energy Resources: Prospects and Priorities*, International Research Center for Energy and Economic Development, Boulder, Colo., 1977.

———, Mihssen Kadhim, and Barry Poulson: *Capital Investment in the Middle East*, Praeger Publishers, New York, 1977.

Mendershausen, Horst: *Coping with the Oil Crisis*, The Johns Hopkins University Press, Baltimore, Md., 1976.

Oppenheim, V. H.: "Why Oil Prices Go Up (1) The Past: We Pushed Them," *Foreign Policy*, no. 25, 1967–1977, pp. 24–57.

Rustow, Dankwart A., and John F. Mugno: *OPEC—Success and Prospects*, New York University Press, New York, 1976.

Sampson, Anthony: *The Seven Sisters*, Hodder and Stoughton, London, 1975.

Tugendhat, Christopher, and Adrian Hamilton: *Oil—The Biggest Business*, Eyre Methuen, London, 1975.

Turner, Louis: "Oil and the North-South Dialogue," *The World Today*, February 1977, pp. 52–61.

———, and James Bedore: "Saudi and Iranian Petrochemicals and Oil Refining: Trade Warfare in the 1980s," *International Affairs*, October 1977, pp. 572–586.

U.S. Congress, Senate Committee on Energy and Natural Resources: *Access to Oil—The United States Relationship with Saudi Arabia and Iran*, 95th Cong., 1st Sess., 1977.

U.S. Congress, Senate Committee on Interior and Insular Affairs: *Geopolitics of Energy*, 95th Cong., 1st Sess., 1977.

Vernon, Raymond: *Storm over the Multinationals*, Harvard University Press, Cambridge, Mass., 1977.

Willrich, Mason, with Joel Darmstadter, et al.: *Energy and World Politics*, The Free Press, New York, 1975.

Index

About the Author

DR. ØYSTEIN NORENG is a professor at the Oslo Institute of Business Administration. Dr. Noreng has worked previously as a counselor in the Planning Department of the Norwegian Ministry of Finance and as Research and Planning Manager in the Marketing Department of Statoil, Norway's state oil company. He has participated in working groups at the Norwegian Institute of International Affairs and he is a member of the Centre d'Études de Politique Etrangère, Paris. He received his master's degree from the University of Oslo and his doctorate in political science from the University of Paris. Dr. Noreng was recipient of a Hoff-Farmand Prize from the University of Oslo in 1975, and in 1977–1978 he was at Stanford University on the Trygve Lie Fulbright Fellowship in Social Sciences. He has contributed many articles to Norwegian and foreign journals on oil and energy policy issues and on international affairs.